Level 4

Science

McGraw, Tondy, Toukonen

Bright Ideas Press, LLC
Cleveland, OH

Simple Solutions Level 4
Science

All rights reserved. No part of this publication may be reproduced or transmitted in any form or by any means, electronic or mechanical, including photocopy, recording, or any information storage or retrieval system. Reproduction of these materials for an entire class, school, or district is prohibited.

Printed in the United States of America

ISBN-13: 978-1-934210-62-8
ISBN-10: 1-934210-62-5

Cover Design: Dan Mazzola
Editor: Kimberly A. Dambrogio
Illustrator: Christopher Backs

Copyright © 2010 by Bright Ideas Press, LLC
Cleveland, Ohio

Note to the Student:

We hope that this program will help you understand Science concepts better than ever. For many of you, it will help you to have a more positive attitude toward learning these topics.

Using this workbook will give you the opportunity to remember topics you have learned in previous grades. By revisiting these topics each day, you will gain confidence in Science.

In order for this program to help you, it is extremely important that you do a lesson every day. It is also important that you ask your teacher for help with the items that you don't understand or that you get wrong on your homework.

We hope that through Simple Solutions and hard work, you discover how satisfying and how much fun Science can be!

Lesson #1

Scientific Method

When scientists have a problem or a question, they use an organized plan called the **scientific method** to conduct a study. This study is called an **investigation**. There are 5 steps for planning and conducting an investigation.

The first step is **Observing and Asking Questions**. During the first step, you use your senses to gather information. You may begin to think of some questions about what you are observing. In addition, you may also discover some things you don't know but would like to find out more about. In the first step, you come up with questions about what you are observing.

The second step of the scientific method is **Forming a Hypothesis**. A **hypothesis** is an educated guess or a possible answer to one of your questions. A hypothesis can be tested to see if it is correct. Always write your hypothesis in a complete sentence.

1. List, in order, the first two steps of the scientific method.

    A)_____

    B)_____

2. What is a hypothesis?

    an experiment   an educated guess   a test

3. During the first step of the scientific method, what do you use to gather information?

      a pen and paper   a hand lens   your senses

Juan and Hannah each planted a bean seed. They used centimeter rulers to measure the bean plants that grew from the seeds.

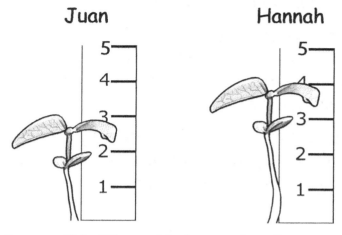

4. About how tall is Hannah's bean plant? _____

5. About how tall is Juan's bean plant? _____

6. How much taller is Hannah's plant than Juan's? _____

7. Which instrument is used to measure how hot or cold something is?

      barometer   balance   thermometer

8 – 10. **Living things need food, water, and air to live.** Look at the chart below. Decide which are **living things** and which are **nonliving things**. Put a ✓ in the correct box for each.

	Living	Nonliving
fog		
butterfly		
rock		

Lesson #2

Scientific Method

The third step of the scientific method is **Planning an Experiment**. An **experiment** is a test that is done to see if your hypothesis is correct or not. When you plan an experiment, you need to describe the steps, list the materials you will need, identify the variables, and decide how you will gather and record your data.

The fourth step of the scientific method is **Conducting an Experiment**. You should follow the steps of the experiment you have planned. Observe carefully and record your information accurately.

The final step of the scientific method is **Drawing Conclusions**. You look at all of the information you have collected. You can make graphs and charts and write a conclusion. Then, decide whether your hypothesis is correct.

1. List the final three steps of the scientific method.

    A)_____

    B)_____

    C)_____

2. Which word means *an educated guess*?

    experiment   hypothesis   conclusion

3. What tool is shown here?

    _____

Simple Solutions© Science											Level 4

4. If a scientist wanted to find out how tall a plant grows each day, the scientist would _____.

   A) put the plant in a sunny window every day
   B) put the plant on a scale and weigh it each day
   C) give the plant a cup of water each day
   D) measure the plant with a ruler each day

5. What could be done to make this kite fly better?

   A) add more color
   B) add another side
   C) add a tail

6. Why do animals need shelter?

   A) for quiet   B) for protection   C) for darkness

7. Write T if the statement is true or F if it is false.

   _____ Living things need food, water, and air to live.

8. Which type of severe weather is shown?

   tornado   snow   thunderstorm

9 – 10. A girl found the skull of an animal. She didn't know what the animal was, but she was sure it preyed on other animals for its food. What could have led her to this conclusion? Answer in complete sentences.

_____

_____

Simple Solutions© Science   Level 4

Lesson #3

1 – 2. Put the steps of the scientific method in the correct order.

    A) Drawing Conclusions   1. _____

    B) Conducting an Experiment   2. _____

    C) Observing and Asking Questions   3. _____

    D) Planning an Experiment   4. _____

    E) Forming a Hypothesis   5. _____

Use the graph to answer the questions below.

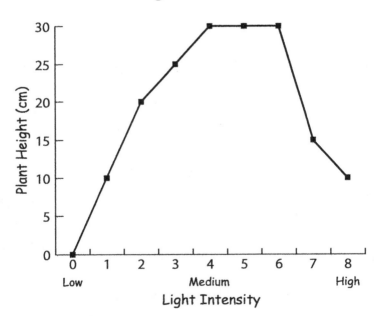

3. What conclusion can be drawn from this graph?

    A) Bean plants grow best in high light.

    B) Bean plants grow best in low light.

    C) Bean plants grow best in medium light.

    D) Bean plants grow the same in any light.

4. Which light intensities caused the bean plants to grow 30 cm? _____

5. What is the name for a behavior an animal knows without being taught?

        hibernate   migrate   instinct

6. What does this tool measure?

   A) wind direction
   B) temperature
   C) rainfall

7. Which is a **living** part of an environment?

        pebble   carrot   water vapor

Use the pictures of bird beaks to answer items 8 and 9.

 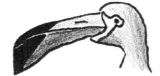 

8. Which beak would be best for sipping nectar from a flower? _____

9. Which beak would be most helpful for a bird that eats insects in the bark of trees? _____

10. Some whales migrate from the North Pole to the equator. Which unit of measurement would be best for measuring this distance?

   A) millimeter   C) meter
   B) kilometer   D) centimeter

Lesson #4

Needs of Plants (Review)

Most plants need **light**, **air**, **water**, and **nutrients**. Nutrients come mostly from the soil.

There are three main parts to every plant: the **root**, the **leaf (leaves)**, and the **stem**. The roots take in water and nutrients from the soil. The stem helps to hold the plant up. The leaves are where the plant makes its food.

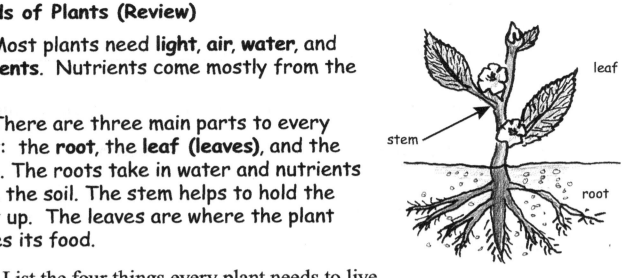

1. List the four things every plant needs to live.

   _____   _____

   _____   _____

2. Which part helps to hold up the plant?

   _____

3. Name the three main parts of a plant.

   _____   _____

   _____

4. Which of these is the first step of the scientific method?

   A) Planning an Experiment

   B) Forming a Hypothesis

   C) Observing and Asking Questions

5. Which color fur will best protect a rabbit from a hawk in a snowy field?

   A) brown   B) gray   C) black   D) white

   Explain your answer choice. _____

   _____

6. What is the first stage of life for many plants?

   root   leaf   seed

7. Which are **nonliving** parts of the environment?

   water   goat   oxygen   water vapor   jellyfish

Seeds contain the food to help a new plant grow. Seeds can be spread to other places. Seeds can be spread by wind, by moving water, by sticking to an animal's fur or to a person's clothes; even birds can carry seeds. Sometimes animals eat fruit, leaving the seeds behind.

8 – 9. Name four ways seeds are spread.

_____   _____

_____   _____

10. Write T if the statement is true or F if it is false.

    _____ The roots are where the plant makes its food.

Lesson #5

1. What are the four basic needs of plants?

    A) light, carbon dioxide, soil, and water
    B) water, light, shelter, oxygen
    C) soil, food, water, carbon dioxide
    D) light, air, water, nutrients

This graph shows the different kinds of birds observed on a class field trip. Use the graph to answer the questions below.

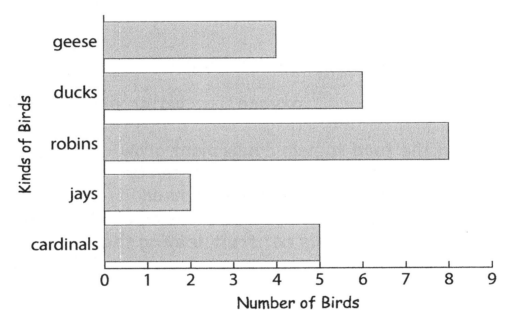

2. How many of the birds seen were robins? _____

3. How many of the birds seen were ducks or geese? _____

4. How many total birds were seen on the field trip? _____

5. Which is **not** something all living things need in order to live?

    water   air   shelter   food

6 – 7. Decide which are **living things** and which are **nonliving things**. Put a ✓ in the correct box for each.

	Living	Nonliving
basket		
root		
wind		
poison ivy		

8. An instrument that measures wind speed is called an **anemometer**. Write the word on the line below.

   _____

9. Your town passes a law that makes it illegal to hunt deer. Which will most likely happen during the following year in your town's forests?

   A) The number of deer will decrease.

   B) The number of insects will increase.

   C) The number of trees will increase.

   D) The number of small plants will decrease.

10. What does a school nurse use to measure body temperature?

   A) a ruler

   B) a balance

   C) a thermometer

   D) a stopwatch

Lesson #6

A **barometer** is a tool that measures air pressure and tells whether the pressure is rising, falling, or remaining steady. A barometer is just as accurate as the weather forecaster on the news. Under most conditions a barometer can forecast weather up to 12 to 24 hours ahead. Changes in pressure readings will be of greater magnitude in winter months than in summer, and will also depend on location and altitude.

1. What is the name of the tool that measures air pressure?

   _____

2. How are roots and stems alike?

   A) Both are green.

   B) Both grow underground.

   C) Both move water through the plant.

3. Name four basic needs of plants.

   _____   _____

   _____   _____

4. Which is **not** a living thing?

   rose bush   soil   squirrel   root

5. In which step of the scientific method do you describe the steps, list the materials you will need, identify the variables, and decide how you will gather and record your data?

   A) Observing and Asking Questions
   B) Forming a Hypothesis
   C) Planning an Experiment
   D) Drawing Conclusions

6 – 8. Match each type of precipitation below with its description in the table.

   hail   snow   rain   sleet

Type of Precipitation	Description
	drops of water fall, temperatures above freezing
	frozen rain; when rain falls through a layer of freezing cold air
	round pieces of ice; rain freezes and falls to a warmer part of the air
	drops of water vapor form ice crystals as they fall

9. Write T if the statement is true or F if it is false.

   _____ Seeds can be spread by wind and water.

10. The _____ are where the plant makes its food.

Lesson #7

Classifying Plants/Vascular

The plant kingdom is divided into two groups. Plants are either **vascular** or **nonvascular**. **Vascular** plants have tube-like structures that transport water from the roots to the stem and then to the leaves. Vascular plants are plants with a stem, leaves, and roots. Vascular plants are divided into 3 smaller groups: **flowering plants**, which make seeds in fruits, **cone-bearing plants**, which make seeds in cones (conifers), and **ferns**, which do not make seeds.

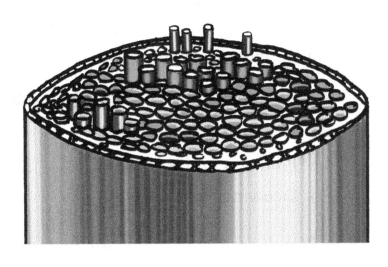

1. The plant kingdom is divided into two groups. Name them.

   _____

   _____

2. Vascular plants are divided into three smaller groups. Name the three groups. One has been done for you.

   <u>flowering plants</u>

   _____

   _____

3. Which plant group does not makes seeds?

   _____

4. Erica wants to measure air pressure. Which instrument should she use?

   A) anemometer
   B) barometer
   C) wind vane

5. Hank thinks eating breakfast every morning will help him stay awake. He has not tested this idea out yet. It is his _____.

   theory   conclusion   hypothesis   experiment

6. Which is a living thing?

   cloud   rock   hail   root

7. Marshal wants to know how long and wide his desk is. He should measure it with a _____.

   balance   thermometer   anemometer   ruler

8. Most plants need nutrients. Where do plants get most of their nutrients?

   the air   water   the sun   the soil

9. Name the three main parts of a plant.

   _____   _____   _____

10. Which of these is **not** a basic need of plants?

    water   fertilizer   nutrients   light   air

Lesson #8

Classifying Plants/Nonvascular

Nonvascular plants absorb water through their surfaces, like a sponge. They do not have any tubes to carry water and food to the parts of the plant. **Moss** is an example of a nonvascular plant. Moss grows where it's moist. Moss can be found growing on trees (picture top), rocks, buildings, and damp patios or sidewalks. Another example of a nonvascular plant is the **liverwort** (picture right). Liverworts look like green ribbons growing across the ground, with each segment forking into two new branches. Liverworts grow in damp forests and along rivers.

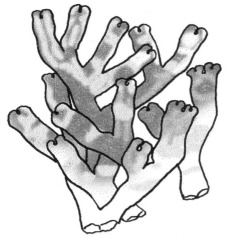

1. How do nonvascular plants absorb water?

   _____

2. Name two examples of nonvascular plants.

   _____

   _____

3. Sleet, rain, snow, and hail are forms of _____.

   evaporation   weather   precipitation

4. The functions of a plant's roots are to support the plant and ____.

   A) make food   C) produce fruit

   B) take in water and nutrients   D) make seeds in cone

A **balance** is a tool that is used to measure the mass of objects. When you place an object in one pan and another object in the other pan, you are able to compare the objects' masses. Mass is measured in grams.

5. A balance measures _____.

    height   distance   mass   temperature

6. Which has more mass, the hummingbirds or the sparrow?

_____

7 – 10. Match each tool below with its description in the table.

microscope   hand lens   balance   thermometer

Tools	Description
	This tool is helpful to see objects that are too small to see with your eyes alone.
	This tool measures how hot or cold something is.
	This tool measures the mass of objects.
	This hand-held tool is used to magnify, or make something look larger.

Lesson #9

1. In which step of the scientific method do you look at all of the information you have collected to see if the results support your hypothesis?

   A) Conducting an Experiment
   B) Observing and Asking Questions
   C) Drawing Conclusions
   D) Forming a Hypothesis

2. The sunrise and sunset times for three days in March are recorded in the chart below.

Date	Sunrise	Sunset
March 9	7:00 a.m.	5:15 p.m.
March 16	6:45 a.m.	5:30 p.m.
March 23	6:35 a.m.	5:40 p.m.

   Which statement is an accurate conclusion based on this information?

   A) There are fewer hours of daylight as the month goes on.
   B) There are more hours of daylight as the month goes on.
   C) The sun rises later as the month goes on.
   D) The sun sets earlier as the month goes on.

3. Write T if the statement is true or F if it is false.

   _____ Vascular plants are plants with a stem, leaves, and roots.

4. A balance measures the _____ of an object.

5. The plant kingdom is divided into two groups. Name one of them.

_____

Use the diagram of a fish tank to answer the questions below.

6. Identify two living things in the diagram.

_____

_____

7. Identify two nonliving things in the diagram.

_____

_____

8. Vascular plants are divided into three smaller groups. Which of these is correct?

   A) mosses, liverworts, and ferns
   B) ferns, flowering plants, and cone-bearing plants
   C) mosses, ferns, and flowering plants

9 – 10. All living things have three basic needs. What are they?

_____   _____

_____

Lesson #10

Fungi

Fungi were once classified as plants, but fungi cannot make food. They absorb food and cannot move around. Fungi absorb nutrients from other living things or from the remains of living things. Fungi look like plants because they grow upright. Some examples of fungi are **mushrooms** and **mold**.

1. Which characteristic of fungi keeps them from being classified as plants?

    A) They don't have flowers.
    B) They don't grow upright.
    C) They cannot make food.

2. Give two examples of fungi.

    _____

    _____

3. Fill in the missing steps of the scientific method.

    1. _____

    2. _Forming a Hypothesis_____

    3. _____

    4. _Conducting an Experiment___

    5. _____

4 – 5. Look at the chart below. Decide which are living things and which are nonliving things. Put a ✓ in the correct box for each.

	Living	Nonliving
pine tree		
snow		
soil		
root		
stone		

6. Two objects of the same mass are going to be placed the same distance from the support of a balance beam. Which diagram shows where the support should be placed so the objects will balance?

    A)   B)   C)   D)

7. What would you use to measure the height of a plant stem?

    balance   thermometer   ruler   weather vane

8. Anything too small to be seen with the eyes alone is _____.

    vascular   fungi   microscopic   nonvascular

9 – 10. Name four ways seeds are spread.

_____ _____

_____ _____

Lesson #11

1. Which organism absorbs nutrients from other living things or from the remains of living things?

    moss   fern   fungi   liverwort

2. Which is **not** true of flowering plants?

    A) They are vascular plants.
    B) They make food in the leaves.
    C) They produce seeds.
    D) They make cones.

The chart below shows temperatures recorded from Monday through Friday during a week in July.

Day	Temperature (°F)
Monday	76
Tuesday	72
Wednesday	70
Thursday	68
Friday	74

3. Which thermometer shows the temperature recorded on Wednesday?

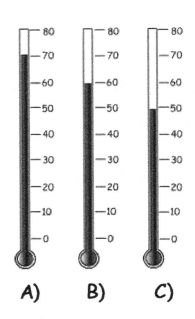

4. Which tool would you use to take a closer look at a cocoon?

    microscope   balance   hand lens   barometer

5. What is a hypothesis?

    an educated guess   an experiment   an observation

6. Maple trees, ferns, and mosses are similar because they are all __.

    bulbs   plants   nonliving   vascular

7. In which step of the scientific method do you use your senses to gather information?

    A) Forming a Hypothesis
    B) Drawing a Conclusion
    C) Observing and Asking Questions

8. A balance measures _____.

    mass   distance   temperature   time

9. Underline two examples of fungi.

    moss   mushroom   fern   poison ivy   mold

10. Which of these is **not** something all plants need to live?

    nutrients   water   soil   air   light

Lesson #12

1. Put a ✓ next to the one statement that is true about nonvascular plants.

   _____ They have stems and roots.

   _____ They have conducting tubes.

   _____ They absorb water directly, like a sponge.

2. Some animals go into a deep sleep for winter. What is this called?

   A) camouflage   B) hibernation   C) migration

In science we use tools to help us observe, measure, or study objects. One of these tools is used to magnify, or make something look larger. It is called a **hand lens**.

3. What do all living things need?

   A) food, water, soil
   B) water, air, shelter
   C) food, air, water

4. Draw your own picture of a hand lens in this box.

5. Fill in the missing steps of the scientific method.

   1. <u>Observing and Asking Questions</u>

   2. _____

   3. <u>Planning an Experiment</u>

   4. _____

   5. <u>Drawing Conclusions</u>

6. Why don't most plants grow on rocks?

       no light   no soil   no water

7. Which part of the plant carries water and nutrients to the leaves?

       roots   stem   flower

8. Write T if the statement is true or F if it is false.

   _____ Vascular plants are divided into flowering, cone-bearing, and ferns.

   _____ A barometer measures wind speed.

9. Which is a **living** part of an environment?

       soil   cloud   root   snow

10. Which tool would you use to look at bacteria in pond water?

    anemometer   microscope   thermometer   barometer

Lesson #13

Photosynthesis

Unlike animals, plants can make their own food. The food that plants make is sugar. The sugar is made by a process called **photosynthesis**. Sugar is created in the green parts of a plant. **Chlorophyll** is the green substance inside leaves. Chlorophyll absorbs sunlight.

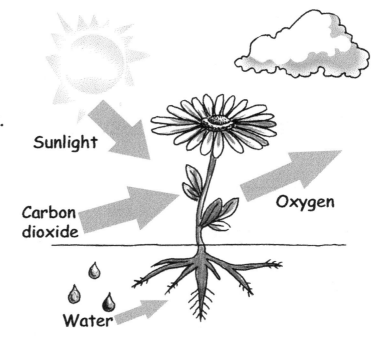

Sunlight provides the energy for photosynthesis to happen.

Plants take a gas called carbon dioxide from the air; they pull water up through their roots, and use sunlight to make sugar. This sugar provides the energy that plants need for growth and health. Plants give off oxygen as a by-product. The plant uses the sugar for energy. Plants use some of this energy right away and the rest is stored in the leaves, stem, and seeds for later use. The equation for photosynthesis is:

carbon dioxide + water → sugar + oxygen

1. Where do plants store the food that they make?

   _____   _____   _____

2. What do plants absorb through their root system?

   _____

Simple Solutions© Science   Level 4

3. Write the formula for photosynthesis on the line.

   _____

4. What is the gas that plants give off during photosynthesis?

   _____

5. The giraffe is warm-blooded. It is a mammal; it is very tall. These are all _____ of the giraffe.

   characteristics   parts   sizes

6. When birds fly south for the winter, what are they doing?

   hibernating   migrating   growing   hiding

7. The plant kingdom is divided into two groups. Underline them.

   moss   vascular   fungi   nonvascular   mushroom

8. Which step of the scientific method comes right after **Observe and Ask Questions**?

   _____

9. An anemometer measures wind _____.

   direction   speed   temperature

10. Write T if the statement is true or F if it is false.

    _____ Wind is an example of a living thing.

Lesson #14

Vertebrates

Animals are divided into two groups —animals with a backbone and animals without a backbone. Those animals with a backbone belong to a group called **vertebrates**. There are 5 groups of vertebrates: mammals, amphibians, birds, fish, and reptiles. **Mammals** are vertebrates that have hair or fur. They use lungs to breathe and they are **warm-blooded**. Warm-blooded animals try to keep the inside of their bodies at a constant temperature. To generate heat, warm-blooded animals convert the food that they eat into energy.

Most mammals do not lay eggs; their babies are born alive. Mammals also produce milk for their young. Here are some examples of mammals: **human, goat, whale, bat,** and **rabbit**.

Amphibians are vertebrates that spend part of their life in the water and part on land. Amphibians are cold-blooded. Examples of amphibians are **frogs, toads,** and **salamanders**. Young amphibians have gills for breathing and tails that help them swim. (A young frog or toad is called a **tadpole**.) As young amphibians grow, their gills disappear and they develop lungs. They also develop legs and have moist skin. Amphibians usually stay close to the water so they can keep their skin moist. They lay their eggs in the water.

tadpole

1. Animals with a backbone are called _____.

2. Animals without a backbone are called _____.

3. Which vertebrate group has hair or fur and their babies are born alive?

   _____

4. A young frog is called a _____.

5. Which of these has the greatest mass?

   a paper clip   a bowling ball   a mouse   a semi truck

6. Put a ✓ next to each statement that is true about vascular plants.

   _____ They absorb water directly, like a sponge.

   _____ They have tubes that transport water.

   _____ They have stems, leaves, and roots.

7. What provides the energy for photosynthesis to happen?

   carbon dioxide   oxygen   sunlight   water

8. What does this picture show?

   vascular plants   fungi   nonvascular plants

9. This tool is used to magnify, or make something look larger. What is it called?

   _____

10. Name the three main parts of a plant.

   _____   _____   _____

Lesson #15

Vertebrates

Birds are vertebrates that have feathers, wings, and two legs. Feathers protect the bird, keep them warm, and help them fly. Like mammals, birds breathe with their lungs. They lay eggs with hard shells. Birds are **warm-blooded**.

Fish are vertebrates that live in water. Most fish are covered with scales that protect them and help them swim. Most fish lay eggs. Fish take in oxygen through their gills. Fish have fins to help them swim. Fish are **cold-blooded**.

Reptiles are vertebrates that have dry, scaly skin and lay eggs on land. Reptiles breathe with their lungs. They are **cold-blooded**. This means that their body temperature changes with their surroundings. Cold-blooded animals cannot maintain their own stable temperature like warm-blooded animals can. The scales on their skin help to protect them from the hot sun. Most reptiles hatch from eggs. Reptiles that spend a lot of time in the water must go to the surface to get the oxygen they need. Examples of reptiles include **alligators**, **snakes**, **turtles**, and **iguanas**.

1. Vertebrates that have feathers, wings, and two legs are called _____.

2. Which two vertebrate groups are warm-blooded?

    reptiles and birds   birds and mammals   fish and birds

3. Vertebrates that are cold-blooded with dry, scaly skin are _____.

4. _____ is the vertebrate group that spends its entire life in water.

5. Choose the correct formula for photosynthesis.

   A) oxygen + water → sugar + carbon dioxide
   B) carbon dioxide + nutrients + water → sugar
   C) carbon dioxide + water → sugar + oxygen

6 – 8. Use the word bank to complete the sentences below.

sugar	water	photosynthesis
sunlight	oxygen	carbon dioxide

During _____, plants take in a gas called _____ from the air; they pull _____ up through their roots, and use _____ to make _____. Plants give off _____ as a by-product.

9. Animals with backbones are called _____.

      invertebrates   vertebrates   vascular

10. What is a tool that measures air pressure?

    anemometer   wind vane   barometer   thermometer

Lesson #16

1. List the five groups of vertebrates.

   _____

   _____

   _____

   _____

   _____

2. An organism without a backbone is called a(n) _____.

   vertebrate   carnivore   invertebrate

3. Write **T** if the statement is true or **F** if it is false.

   _____ Fish are warm-blooded.

   _____ Tadpoles breathe through gills.

4. Which instrument is shown?

   anemometer   microscope   hand lens

5. Which of these is the final step of the scientific method?

   A) Planning an Experiment
   B) Forming a Hypothesis
   C) Observing and Asking Questions
   D) Drawing Conclusions

6. List the parts of a plant.

   _____

   _____

   _____

7. The plant kingdom is divided into two groups. What are they?

   A) mushrooms and molds
   B) ferns and cone-bearing plants
   C) vascular and nonvascular plants
   D) flowering plants and cone-bearing plants

8. What do these vertebrate groups have in common?

   amphibian   fish   reptile

   A) They all breathe with gills.
   B) They all produce milk for their young.
   C) They are all cold-blooded.
   D) They are all warm-blooded.

9. _____ is a by-product of photosynthesis.

   Carbon dioxide   Chlorophyll   Oxygen

10. What do you call the part of the plant that grows above ground and helps hold the plant up?

    root   leaf   stem   flower

Lesson #17

Invertebrates

Animals without a backbone belong to a group called invertebrates. There are over a million kinds of invertebrates, and many of them live in the oceans. Insects are also invertebrates, and are, in fact, the largest group of invertebrates. Most insects have 3 body parts (head, thorax, and abdomen) and six legs. They have hard outer coverings. Many insects have two pair of wings and a pair of antennae. Spiders are also invertebrates. Spiders look like insects, but they are not insects. 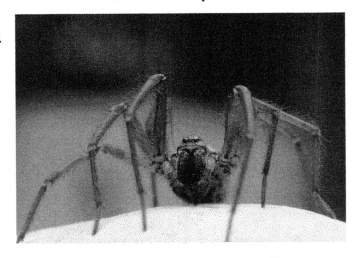 Spiders have eight legs, 2 body parts, and an outer covering. All spiders make silk. Other examples of invertebrates are the worm, sponge, jellyfish, snail, squid, crab, scorpion, shrimp, octopus, starfish, lobster, and butterfly.

1. _____ are the largest group of invertebrates.

2. List three examples of an invertebrate.

   _____

   _____

   _____

3. An organism with a backbone is called a(n) _____.

   vertebrate   fungus   invertebrate

4. Fill in the missing information for spiders and insects.

	Number of Legs	Number of Body Parts
Insect		
Spider		

5. Which word means *an educated guess*?

   hypothesis   investigation   observation

6. What do spiders and insects have in common?

   number of legs   outer body covering   wings

7. Some animals go into a deep sleep for winter. What is this called?

   camouflage   hibernation   migration

8. Write T if the statement is true or F if it is false.

   _____ Chlorophyll is the green substance inside leaves.

   _____ A mushroom is a fungus.

9 – 10. What are the four basic needs of animals?

_____   _____

_____   _____

Lesson #18

1. What is the gas that plants give off during photosynthesis?

   _____

2. Match each type of precipitation below with its description.

   ____ sleet   A) drops of water fall, temperatures above freezing

   ____ snow   B) round pieces of ice; rain freezes and falls to a warmer part of the air

   ____ rain   C) frozen rain; rain falls through a layer of freezing cold air

   ____ hail   D) drops of water vapor form ice crystals as they fall

3. Seeds can be spread by wind, by water, and by _____.

   gills   animals   photosynthesis

4. Jenna reaches her hand into a bag filled with smooth objects. Jenna feels the objects, but cannot look into the bag. Which property of the objects can she most likely identify?

   A) color   B) shape   C) mass

5. Which part of a plant produces seeds?

   A) leaves
   B) stem
   C) flower
   D) roots

6. What word means *the colors and patterns an animal uses to disguise itself*?

    instinct   camouflage   migrate

7. To which group do birds and fish belong?

    A) reptiles
    B) amphibians
    C) mammals
    D) vertebrates

Use the Life Span chart to answer the questions below.

Animal	Life Span
Mouse	2 years
Hummingbird	9 years
Snapping Turtle	35 years
House fly	21 days

8. Which animal has the longest life span?

    _____

9. Which animal has the shortest life span?

    _____

10. How much longer do snapping turtles live than hummingbirds?

    _____

Lesson #19

1. Which of these has the greatest mass?

    a rubber band   a basketball   a couch   a hamster

Unlike animals, plants cannot run away from their enemies. They need other ways to protect themselves. Some plants give off an odor, some have thorns, some can make an animal sick, and some plants can even kill anything that eats them.

2 – 3. Use the words below to complete the sentences. (One word will not be used.)

    odor   needles   thorns   poisons   rash

Rosebushes have _____ on their branches. Some

plants contain _____ that can causes sickness if

they are eaten. Cacti protect themselves with _____.

Marigolds give off a(n) _____ that keeps insects away.

4. The fish, the cat, and the bird are all alike in many ways. One way is that they all have _____.

    A) lungs
    B) wings
    C) legs
    D) backbones

5. A young frog is called a _____.

Simple Solutions© Science   Level 4

An animal that hunts another animal for food is called a **predator**. A wolf is a predator and so is a hawk. An animal that is hunted for food is called **prey**. Antelope are prey for the wolf, and chipmunks are prey for the hawk. Some animals are both predators and prey. If a small bird eats insects or worms, the bird is a predator. If a hawk eats the small bird, then the bird becomes the prey, and the hawk is the predator.

6. An animal that is hunted for food by other animals is called _____.

    predator   vertebrate   prey   warm-blooded

7. A wolf hunts and eats an antelope. In this situation, the wolf is the _____.

    prey   instinct   predator   invertebrate

8. All of the following are needed for a fish to live in an aquarium except _____.

    oxygen   food   rocks   water

9. What do these animals have in common?

    alligator   frog   catfish   snake

   A) They are all invertebrates.
   B) They are all reptiles.
   C) They are all amphibians.
   D) They are all cold-blooded.

10. What does vascular mean?

    square-shaped   with tubes   with cones   having flowers

Lesson #20

What is a Food Chain?

A food chain shows the path of food from one living thing to another. <u>No matter which organisms are part of the food chain, the sun is always the first link.</u> After the sun, all food chains begin with a producer. A **producer** produces its own food. Plants and vegetables are producers. All energy originally comes from the sun, and plants use that energy to make food. All animals depend on plants for their energy. Some animals eat plants to get energy, and that is how they live and grow; others eat other animals. A **consumer** is an organism that eats other living things (plants or animals) in order to get energy. Animals are consumers. They get their energy by eating animals or by eating animals that eat plants.

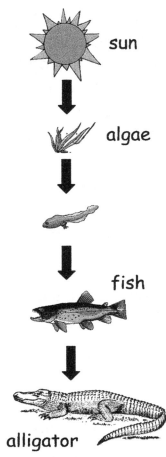

In this food chain, the algae get their energy from the sun. The tadpole eats the algae for its food, the fish eats the tadpole, and the alligator eats the fish. The algae are the producers, and the tadpole, fish, and alligator are all consumers.

1. What shows the path of food from one living thing to another?

   _____

2. All food chains begin with a _____.

3. What do you call an organism that eats other living things in order to get energy?

         a producer   a consumer   a food chain

Simple Solutions© Science   Level 4

4. An owl hunts and eats a mouse. The mouse is the _____.

   predator   producer   prey

5 – 6. Match each animal group to its definition.

   _____ Fish   A) has hair or fur; feeds young with milk

   _____ Reptile   B) two legs; wings and feathers

   _____ Bird   C) moist-skin; lives near water

   _____ Amphibian   D) dry, scaly animal that lays eggs

   _____ Mammal   E) lives whole life in water; breathes with gills

7. List the three needs of all living things.

   _____   _____

   _____

8. Which of these is an invertebrate?

   goose   lizard   spider   rabbit

9. What is the green substance inside leaves?

   sugar   chlorophyll   photosynthesis   tubes

10. What is one important way scientists group animals?

   A) with or without a backbone   B) size   C) height

Simple Solutions© Science   Level 4

Lesson #21

1 – 2. Use the Venn diagram below to compare and contrast insects and spiders. List two similarities and two differences. One similarity has been given.

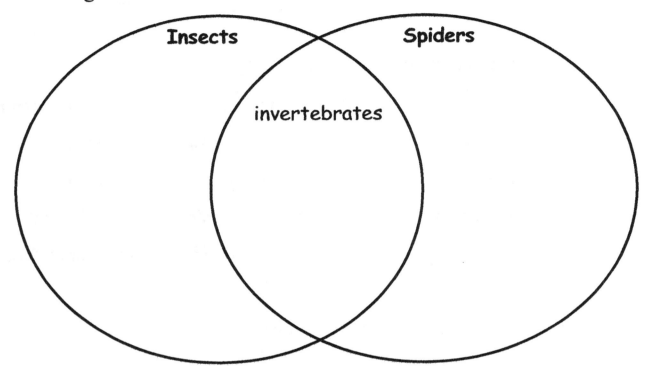

3. What word means *to go into a sleep-like state* during the winter months?

   camouflage   hibernate   migrate   instinct

4. A plant produces _____ through photosynthesis and then uses it for energy.

   oxygen   sugar   water   carbon dioxide

5. Both plants and animals need air to survive. What part of the air do animals use and what part of the air do plants use?

   _____

   _____

6. Which of these is an example of a nonvascular plant?

   A) oak tree
   B) fern
   C) moss
   D) mold

7. Arrange the steps of the scientific method in the correct order.

   A) Planning an Experiment   1) _____
   B) Drawing Conclusions   2) _____
   C) Conducting an Experiment   3) _____
   D) Forming a Hypothesis   4) _____
   E) Observing and Asking Questions 5) _____

8. The path of food from one living thing to another is called the _____.

   photosynthesis   food chain   migration

9. Which word means *an educated guess*?

   hypothesis   experiment   investigation

10. A wind vane measures _____.

    wind speed   wind direction   wind temperature

Lesson #22

1 – 2. Decide which vertebrate group each animal belongs to. Next to each animal name, write how it reproduces by writing either **lays eggs** or **born alive**. Write **warm-blooded** or **cold-blooded**.

Animal	Reproduces	Warm/Cold-Blooded
salmon		
snake		
cat		
chipmunk		
owl		

3. Most plants get their nutrients directly from _____.

   the sun   the air   the soil   other plants

4. Which are invertebrates?

   bat   shrimp   butterfly   robin   dragonfly

5. Name three ways seeds can spread.

   _____

   _____

   _____

Simple Solutions© Science

Level 4

An **herbivore** is an animal that eats only plants. Some examples of herbivores are deer, sheep, horses, and cows.

A **carnivore** is an animal that eats only other animals. Examples of carnivores are tigers, foxes, alligators, and coyotes.

An **omnivore** eats both plants and animals. Some examples of omnivores include chickens, raccoons, chimpanzees, and bears.

6. A(n) _____ eats other animals.

7. A(n) _____ eats only plants.

8. A(n) _____ eats both plants and other animals.

9. Which has more mass, the sparrow or the chicken?

    _____

10. Put a ✓ next to each statement that is true about photosynthesis.

    _____ Carbon dioxide is a by-product of photosynthesis.

    _____ Sunlight provides the energy needed for photosynthesis to happen.

Lesson #23

1. Which of the following is a trait of reptiles?

    A) hair
    B) gills
    C) scales
    D) feathers

2. What are the four basic needs of plants?

    A) light, air, water, nutrients
    B) soil, food, water, carbon dioxide
    C) water, light, shelter, oxygen
    D) light, carbon dioxide, soil, and water

3. Which is a nonliving part of the environment?

    snail   mushroom   carbon dioxide   weeds

4. What are two traits of birds?

    _____

    _____

5. What are the two main groups of plants?

    A) vertebrates and invertebrates
    B) molds and mushrooms
    C) vascular and nonvascular
    D) producers and consumers

Simple Solutions© Science — Level 4

Use the food chain below to answer questions 6 – 8.

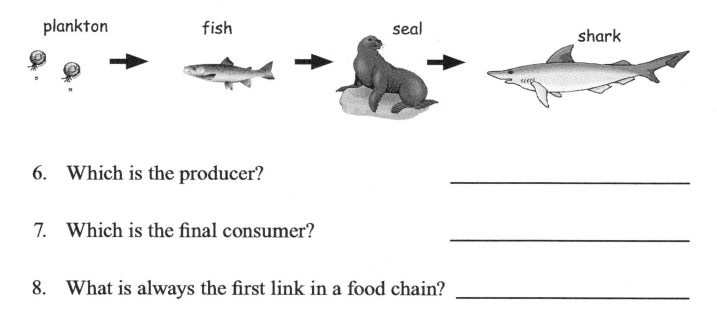

plankton → fish → seal → shark

6. Which is the producer? _____

7. Which is the final consumer? _____

8. What is always the first link in a food chain? _____

Animals and insects grow and change during their lifetimes. In some animals, this series of changes in appearance from birth to adulthood is called **metamorphosis**. Butterflies, moths, frogs, and toads go through **complete metamorphosis**. When a butterfly goes through metamorphosis, the butterfly changes the way it looks during its life cycle. Its life cycle begins with an **egg** laid on a leaf. The egg hatches into a caterpillar. This stage is called **larva**. The larva eats leaves. Next the caterpillar becomes a **pupa**. In this stage the caterpillar is wrapped in a cocoon. It does not eat or move. When it's time, an **adult** comes out of the cocoon. Now it is a butterfly.

9. The series of changes from birth to adulthood is called _____.

10. What do you call the part of the plant that grows underground and takes water and nutrients from the soil?

　　　　leaf　　　　root　　　　stem　　　　seed

Lesson #24

1. In which stage of metamorphosis is the organism wrapped in a cocoon?

    larva   egg   pupa

2. In which stage of metamorphosis is the organism a caterpillar?

    larva   egg   pupa

3. What do you call an organism that makes its own food?

    a producer   a consumer   a food chain

4. The path of food from one living thing to another is called the _____.

    photosynthesis   food chain   migration

5. For each animal in the chart below, decide whether it is an **herbivore**, a **carnivore**, or an **omnivore**, and put a ✓ in the correct column.

	Herbivore	Carnivore	Omnivore
coyote			
sheep			
raccoon			
chimpanzee			

Use the food chain to answer the questions below.

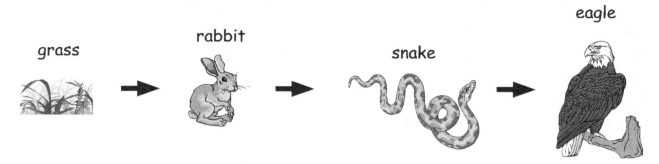

6. What relationship do eagles have with snakes?

   A) Snakes eat eagles.
   B) Eagles eat snakes.
   C) Eagles and snakes both eat grass.

7. In the above food chain, which is the producer?

   the rabbit   the eagle   the grass   the snake

8. Where does the grass get the energy to make its food?

   the soil   the sun   the air   the rain

9. Which of these is the second step of the scientific method?

   A) Drawing Conclusions
   B) Planning an Experiment
   C) Forming a Hypothesis
   D) Conducting an Experiment

10. Which group of plants absorbs water through its surfaces?

    vascular   vertebrate   nonvascular   invertebrate

Lesson #25

Predicting Weather/Air Masses

The weather can be sunny and hot one day and cold and snowing the next. The moving of air masses is what causes changes in the weather. What is an air mass? Simply stated, an **air mass** is a large body of air. Air masses that form over land are dry, and air masses that form over water have a lot of moisture in them. In the United States, cold air masses come from the North and warm air masses come from the South.

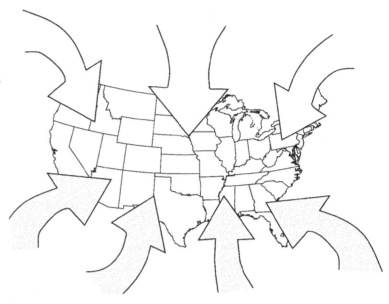

The temperature and moisture levels determine the kind of weather the air mass will bring. A cold, wet air mass can bring snow, but a cold, dry air mass can bring little or no precipitation. Warm air masses with a lot of moisture usually bring precipitation. Warm, dry air masses bring the warm temperatures but little or no precipitation. Air masses usually stay separate from each other as they move about. Weather patterns generally move from west to east in the United States.

1. What is an air mass?

   _____

2. In the United States, weather patterns generally move from

   _____ to _____ .

3. In the United States, cold air masses come from the _____

   and warm air masses come from the _____ .

4. Match each stage of butterfly metamorphosis with its definition.

    _____ larva   A) the caterpillar is wrapped in a cocoon

    _____ adult   B) the life cycle begins here

    _____ pupa   C) the egg hatches into a caterpillar

    _____ egg   D) a butterfly comes out of the cocoon

5. Some consumers get energy by _____.

    eating a producer   going through photosynthesis   migrating

6. A balance measures _____.

    size   temperature   mass   speed

7. What is the name for a living thing that cannot make its own food?

    producer   consumer   fern

8. If a bird flies south each winter, what is it doing?

    adapting   hibernating   migrating

9. Plants make their own food through a process called _____.

10. Animals with a backbone are called _____.

Simple Solutions© Science   Level 4

Lesson #26

1. What makes all animals consumers?

   A) Animals need shelter to survive.

   B) Animals make their own food.

   C) Animals need more energy than plants.

   D) All animals eat plants or other animals.

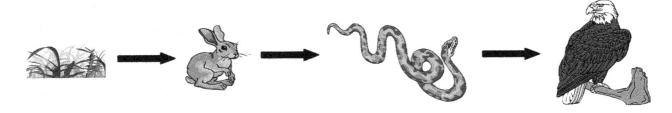

2 – 3. Look at the food chain above. People are beginning to build homes in an area that has a large population of snakes. The builders are putting out poison to kill the snakes. How will this affect the rabbit population in this area?

   _____

   _____

4. What are three traits of fish?

   _____

   _____

   _____

5. What does vascular mean?

   without tubes   with thorns   with stems   with tubes

Simple Solutions© Science Level 4

Use the *Fact Cards* in the Help Pages to fill in the information on the honeybee, the tarantula, and the praying mantis and answer the questions below.

	Diet	Life Span	Defense
Honeybee (Worker & Drone only)			
Tarantula			
Praying Mantis			

6. Which two eat other insects and spiders?

   _____

7. Which has the shortest life span? _____

8. What defense does a praying mantis use? _____

9. Plants make food in a process called photosynthesis. What does a green plant use to make food during photosynthesis?

   A) soil, fertilizer, and water

   B) seeds, water, and energy from sunlight

   C) water, carbon dioxide, and energy from sunlight

   D) oxygen, energy from sunlight, and carbon dioxide

10. An anemometer measures wind _____.

Lesson #27

Predicting Weather/Fronts

When two different large air masses meet, a **front** is formed. Fronts can be very big, or they can be smaller, affecting only a small area. Fronts usually bring changes in the weather. There are two main types of fronts: cold fronts and warm fronts.

A **cold front** is made of cold, dense air. A cold front shows up on a weather map as a blue line with triangles on it. The triangles point in the direction the front is moving. A cold front forms where a cold air mass moves under a warm air mass. This causes the warm air to move up and begin to cool.

A **warm front** is shown on a weather map as a red line with half circles on it. A warm front is formed when a large mass of warm air takes over the cooler air mass. Warm fronts usually move slowly and bring steady rain, rather than thunderstorms. Fronts that stay in one place for many days are called **stationary fronts**.

Cold Front

Warm Front

1. When two different large air masses meet, a _____ is formed.

2. Name two main types of fronts.

   _____   _____

3. Draw the symbol for a cold front in the box.

Simple Solutions© Science Level 4

4. Draw a symbol for a warm front in the box.

5. All animals are (producers / consumers) because they eat plants or other animals.

6. Mold is an example of a _____.

   fern   cone   fungus   vascular

7. Which are nonliving parts of the environment?

   soil   mushroom   helium   snow   crab

8. Write T if the statement is true or F if it is false.

   _____ An air mass is a large body of air.

   _____ An herbivore eats both plants and other animals.

9. Which of these is a producer?

   deer   corn plant   sand   sun

10. An insect called a walking stick looks almost like a twig on a tree. This is an example of _____.

    instinct   migration   camouflage   hibernation

Lesson #28

Weather Maps

A **weather** map helps you to know what the weather is over a specific geographic area at a specific time. All of the information for a weather map is collected at weather stations all over the country. This information is reported to the National Weather Service. Weather maps use symbols to show the weather. A sun symbol means it's sunny in an area. A cloud with rain means it is raining in an area. Weather maps also show fronts. The symbol for a warm front is a red line with half circles along it, and a cold front is a blue line with triangles. Many weather maps also show the temperature. In the United States, the temperature is given in degrees Fahrenheit. Almost every other country gives the temperature in degrees Celsius. Other information given on a weather map includes wind speed, wind direction, air pressure, and the high and low temperature for that day.

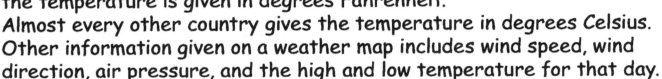

1. What helps you to know what weather is over a specific geographic area at a specific time?

   _____

2. In the United States, which unit is used to measure how warm or cool the air is?

   A) kilograms   C) degrees Fahrenheit

   B) centimeters   D) grams

3. Which type of front does this symbol show?

   _____

Simple Solutions© Science Level 4

4. What is the role of a producer in a food chain?

    A) eat decayed animals   C) eat plants

    B) make food   D) eat other animals

5. Which of the following do herbivores eat?

    omnivores   consumers   predators   producers

6. Which step of the scientific method comes right after **Conduct an Experiment**?

    _____

7. _____ is a by-product of photosynthesis.

8. Which tool would you use to take a closer look at a skin sample?

    balance   thermometer   microscope   barometer

9. Write T if the statement is true or F if it is false.

    _____ Insects are the largest group of invertebrates.

10. Which is **not** a reason feathers are important to birds?

    A) They keep them warm.
    B) They help them fly.
    C) They help them build nests.
    D) They protect them.

Simple Solutions© Science   Level 4

Lesson #29

1. Label the stages of the metamorphosis of a butterfly. (See Help Pages.)

    pupa   adult   larva   egg

    A) _____   C) _____

    B) _____   D) _____

2. Which type of front does this symbol show?

    _____

All animals need **food, water, air,** and **shelter**. Animals must get their food from plants or from other animals. Most animals get the water they need by drinking it or from the foods they eat. Animals need the oxygen from the air. Animals need shelter to help protect them from the weather and from other animals.

3. Animals have four basic needs. List them.

    _____   _____

    _____   _____

4. What are two traits of mammals?

    _____

    _____

Simple Solutions© Science   Level 4

5. Choose the correct formula for photosynthesis.

   A) carbon dioxide + nutrients + water → sugar
   B) carbon dioxide + water → sugar + oxygen
   C) oxygen + water → sugar + carbon dioxide

6. What is the name of the tool that measures air pressure?

   anemometer   barometer   balance   wind vane

Use the graphic organizer to answer the questions below.

Invertebrate	Habitat	Diet	Life Span
octopus	dens, crevices on the sea floor, or in holes under large rocks	small crabs, scallops, snails, fish, turtles, shrimp	about 2 years
snail	gardens, ponds, sea	plants, decaying plants, fruits, and young plant bark	5 – 10 years
lobster	in and around sea weed and rocky habitats	crabs, clams, snails, sea stars, and mussels	50 years

7. Which invertebrates eat snails? _____

8. Which invertebrate has the shortest life span?

   _____

9 – 10. Where does an octopus live?

   _____

Lesson #30

Clouds

Stratus clouds are uniform grayish clouds that often cover the entire sky. They look like a sheet or layer of clouds. Usually no precipitation falls from stratus clouds, but sometimes they may drizzle. When a thick fog lifts, the resulting clouds are low stratus.

Cirrus clouds are thin, wispy clouds blown by high winds into long streamers. Cirrus clouds look like wisps of hair. They are considered "high clouds," forming above 20,000 ft. Cirrus clouds usually move across the sky from west to east. They usually mean fair to pleasant weather.

Cumulus clouds are puffy clouds that sometimes look like pieces of floating cotton. Cumulus clouds are probably the most recognized clouds of all the cloud types. Cumulus clouds are also partly responsible for creating cold front systems.

Cumulonimbus clouds are larger and are more like tall towers than regular cumulus clouds. Fair weather cumulus clouds can form into cumulonimbus clouds in the right conditions. Cumulonimbus clouds are associated with powerful thunderstorms. Snow, rain, hail, lightning, thunder, and sometimes tornadoes can accompany cumulonimbus clouds.

1. Which clouds are puffy and look like pieces of cotton?

   _____

2. Which type of cloud is thin and wispy? _____

3. Which clouds are associated with powerful thunderstorms?

   _____

4. Which clouds are grayish and often cover the entire sky?

   _____

5. Animals without backbones are called _____.

6. Which tool measures how hot or cold something is?

   _____

7. Which of these would a carnivore eat?

   daisy   lettuce   worm   weeds

8. Which list correctly shows how energy moves in a food chain?

   A) carrot ⟶ rabbit ⟶ human ⟶ sun
   B) sun ⟶ carrot ⟶ rabbit ⟶ human
   C) human ⟶ rabbit ⟶ carrot ⟶ sun
   D) rabbit ⟶ carrot ⟶ sun ⟶ human

9. The rabbit in the food chain above gets its energy directly from _____.

   being eaten by humans   eating carrots   the sun

10. Look at the correct food chain in item 8. Which is the producer?

    the human   the rabbit   the sun   the carrot

Lesson #31

1. Animals have four basic needs. List them.

   _____   _____

   _____   _____

2. Match each type of front with its definition.

   _____ warm front   A) stays in one place for many days

   _____ cold front   B) shown on weather maps as a red line with half circles on it

   _____ stationary front   C) shown on weather maps as a blue line with triangles on it

3. What is the gas that plants give off during photosynthesis?

   _____

4. Write T if the statement is true or F if it is false.

   _____ Reptiles have dry, scaly skin.

   _____ Vascular plants have tube-like structures that transport water from the roots to the stem and to the leaves.

5. A robin eats worms and insects. It also eats berries. To which group does the robin belong?

   carnivore   herbivore   omnivore   nonvascular

Simple Solutions© Science — Level 4

6. Fill in the missing steps of the scientific method.

   1. __Observing and Asking Questions__

   2. __Forming a Hypothesis__

   3. _____

   4. _____

   5. __Drawing a Conclusion__

7. Which word means *an educated guess*?

   hypothesis   observation   investigation

8. In which stage of metamorphosis is the organism a caterpillar?

   larva   egg   pupa

9. Which of these is an example of a fungus?

   A) elm tree   C) moss

   B) fern   D) mushroom

10. Which cloud is shown here?

    stratus   cumulus

    cirrus   cumulonimbus

Lesson #32

1. Which of these is a consumer?

    fern   moss   squirrel   daffodil

2. Herbivores and omnivores both eat _____.

    other animals   insects   plants

3. Which clouds are associated with powerful thunderstorms?

    cirrus   stratus   cumulonimbus   cumulus

4. The growth and changes in appearance in some organisms is called _____.

    nonvascular   metamorphosis   invertebrates

5 – 6. Read each phrase below. If the phrase describes an insect, write the word **insect** on the line. If it describes a spider, write the word **spider** on the line. If the phrase describes both, write **both** on the line. (See Lesson #17.)

    A) the largest group of invertebrates   _____

    B) makes silk   _____

    C) has 8 legs and 2 body parts   _____

    D) has an outer covering   _____

    E) has 6 legs and 3 body parts   _____

    F) has 2 pair of wings and a pair of antennae   _____

7. The symbol for which type of front is shown?

   _____

8. A plant produces _____ through photosynthesis and then uses it for energy.

   chlorophyll   sugar   water   carbon dioxide

Use the food chain to answer the question below.

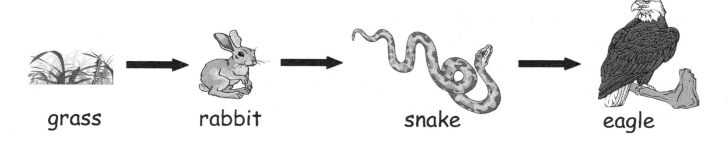

grass   rabbit   snake   eagle

9. Which two animals in the above food chain are carnivores?

   _____   _____

   A **fossil** is the imprint or remains of something that lived long ago. A skeleton or leaf imprint are examples. Most fossils form in sedimentary rock. Fossils help us to learn about plants and animals of the past.

10. The remains of something that lived long ago is called a _____.

Lesson #33

Volcanoes

When pressure from molten rock beneath the Earth's surface becomes too great, the rock, along with lava or gases, escapes through a vent in the crust of the earth. **Volcano** is the term given to both the vent and the cone-like mountain left by the overflow of the erupted lava, rock, and ash. There are different types of volcanoes. **A composite volcano** is made of layers of lava, rock, and ash. Composite volcanoes tend to have steep peaks and are often explosive when they erupt. **A shield volcano** erupts more slowly, and lava flows steadily down the sides. Hawaii has shield volcanoes. A third type of volcano is called a **cinder cone volcano**. Cinder cones are small and have steep sides. This type of volcano sends cinders, ash, and chunks of rock into the air and down the slopes.

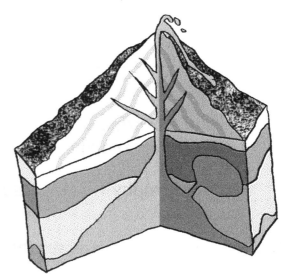

1. The term given to both the vent and the cone-like mountain left by the overflow of erupted lava is _____.

2. Name the three types of volcanoes.

   _____

   _____

   _____

3. What helps us to learn about plants and animals of the past?

   clouds   footprints   fossils   weather

Simple Solutions© Science Level 4

4. Which word means *an educated guess*?

   investigation   conclusion   hypothesis

5. An example of a mammal is a _____.

   crow   whale   lizard   toad

6. Plants make their own food through a process called _____.

Use the Food Chain below to answer the questions below.

7. Which is a producer? _____

8. Which is an herbivore? _____

9. The illustration to the right is an example of what?

   a drawing   a bone   a fossil

10. What is the green substance inside leaves?

    sugar   chlorophyll   photosynthesis   tubes

Lesson #34

Earthquakes

An **earthquake** is a tremor, or shaking of the earth's surface, usually caused by movement of rock in the crust. Earthquakes happen when parts of Earth's crust shift. Small earthquakes happen with movements of just millimeters. Big earthquakes occur with movement of about a meter or more. Most earthquakes occur along **faults**. A fault is a break in the crust, where rock moves.

This ground movement (pictured above) is measured with a **seismograph**. A seismograph (pictured below) shows the movement of the Earth's surface during an earthquake. In 1935, Charles Richter, a **seismologist** (a person who studies earthquakes), developed a scale that measures the size of these movements. The Richter scale rates earth tremors on a scale from 1 to 9, with 9 being the most powerful. A quake that is higher than 4.5 can cause damage to stone buildings. A very severe earthquake would rate a 7 or above.

1. A person who studies earthquakes is called a _____.

2. Who developed a scale to measure the magnitude of Earth's tremors?

   _____

3. _____ is the by-product of photosynthesis.

Simple Solutions© Science   Level 4

4. A break in the crust where rock moves is called a _____.

    volcano   shield   fault   fossil

5. Choose the correct formula for photosynthesis.

    A) carbon dioxide + nutrients + water → sugar
    B) oxygen + water → sugar + carbon dioxide
    C) carbon dioxide + water → sugar + oxygen

6. What do you call a large body of air?

    wind   storm   air mass   stationary front

7. Which are the two main groups of plants?

    A) vertebrates and invertebrates
    B) vascular and nonvascular
    C) ferns and fungus

8. Write T if the statement is true or F if it is false.

    _____ All living things need food, water, and shelter to live.

9. What do you call the part of the plant that grows underground and takes water and nutrients from the soil?

    stem   root   leaf   flower

10. The _____ is always the first link in any food chain.

Lesson #35

Glaciers

In some places on Earth, the snowfall is high and it piles up year after year. As it thickens, it turns to ice. If this mass of ice starts to move downhill, it becomes a glacier. **A glacier** is a large body of ice. This body of ice usually moves slowly down a slope or valley. There are two main types of glaciers: alpine glaciers and ice sheets.

Alpine glaciers flow down mountain valleys. The ice scrapes the floor and the sides of the mountain valley as it moves. This causes the valley to widen, taking on a U shape. A second type of glacier is known as an **ice sheet**. Ice sheets are huge glaciers that can be found in Greenland and Antarctica. Long ago, ice sheets covered much of the Earth.

1. A large body of ice that moves slowly down a slope is called

   a _____.

2. Which type of glacier can be found in Greenland and Antarctica?

   alpine glacier   U shape   ice sheet

3. An instrument that shows the movement of the Earth's surface during an earthquake is called a _____.

   barometer   anemometer   seismograph

Simple Solutions© Science   Level 4

4. Some examples of fungi are _____.

   A) ferns and mosses
   B) molds and mushrooms
   C) flowering plants and molds

5. Which clouds are grayish and often cover the entire sky?

   cumulus   stratus   cirrus   cumulonimbus

6. Name two forms of precipitation.

   _____

   _____

7. A balance measures _____.

   temperature   mass   distance   height

8. Which are **nonliving** parts of the environment?

   soil   grass   carbon dioxide   rat   fog

9. A hawk hunts and eats a sea gull. In this situation, the hawk is the _____.

   predator   producer   prey

10. Most earthquakes occur along _____.

    rocks   hills   rivers   faults

Lesson #36

Landslides

Landslides are a serious problem for people who live in areas with many large hills and mountains. Landslides have affected almost every state in the United States. A **landslide** is the sliding of a mass of loosened rock or earth down a hillside or slope. When a landslide occurs, it can bring water, sand, mud, boulders, trees, and other materials down the hill until the landslide reaches flatter ground. This debris then spreads over large areas, destroying homes, cars, trees, and everything in its path. Landslides can happen more frequently after periods of heavy rain or large amounts of melting snow.

1. The sliding of loose rock or soil down a hillside or slope is called a _____.

2. Which type of glacier covered much of the Earth many years ago?

    ice sheet   U shape   alpine glacier

3. A (cold / warm) front usually moves slowly and brings steady rain, rather than thunderstorms.

4. Animals with backbones are called _____.

5. What three things are needed for photosynthesis to occur?

   A) oxygen, carbon dioxide, and sunlight
   B) carbon dioxide, soil, and sunlight
   C) sunlight, water, and carbon dioxide
   D) water, sunlight, and oxygen

6. Which has more mass, the parrot or the chicken?

   _____

7. In which stage of metamorphosis is a caterpillar?

         larva   egg   pupa

8. Put a ✓ next to each statement that is true about vascular plants.

   _____ Vascular plants have tube-like structures that transport water.

   _____ Fungi are types of vascular plants.

   _____ Vascular plants have leaves, roots, and stems.

9. What word means *the colors and patterns an animal uses to disguise itself*?

       hibernate   migrate   camouflage

10. Which animal group has moist skin and lives near water?

       amphibian   reptile   bird   mammal

Lesson #37

Crossword Puzzle

Word Bank:

vascular	stratus	fault
photosynthesis	front	fossil
glacier	chlorophyll	hypothesis
omnivore	producer	invertebrate

Across

3. A break in the crust, where rock moves
4. This is formed when two different large air masses meet
6. An educated guess
7. The process plants use to make sugar
10. A large body of ice that moves down a slope or valley
11. Plants that have tube-like structures that transport water from the roots to the stem and to the leaves
12. An animal that eats both plants and other animals

Down

1. The green substance inside of leaves
2. An animal without a backbone
5. The remains of something that lived long ago such as a skeleton or leaf imprint
8. Uniform grayish clouds that look like a sheet or layer
9. A living thing that can make its own food

Simple Solutions© Science — Level 4

Use the word bank and the clues on the previous page to complete this puzzle.

Simple Solutions© Science   Level 4

Lesson #38

1. Arrange the steps of the scientific method in the correct order.

   A) Conducting an Experiment   1) _____

   B) Observing and Asking Questions   2) _____

   C) Planning an Experiment   3) _____

   D) Drawing Conclusions   4) _____

   E) Forming a Hypothesis   5) _____

2. A large body of air is called a(n) _____.

   front   barometer   thunderstorm   air mass

Look at the weather map and use it to answer questions 3 and 4.

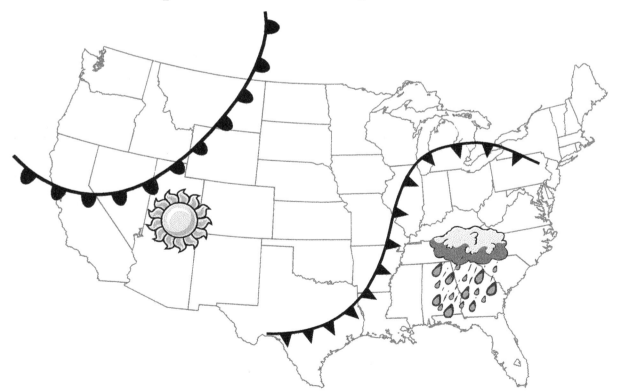

3. Which part of the country is experiencing a warm front?

   the northwest   the Midwest   the northeast

Simple Solutions© Science   Level 4

4. According to the map, which state is having sunny conditions?

    Utah   Tennessee   Kentucky   Georgia

5. What do you call an organism that makes its own food?

    consumer   producer   carnivore

6 – 7. Use the Venn diagram below to compare and contrast reptiles and amphibians. List **three similarities** and **one difference**.

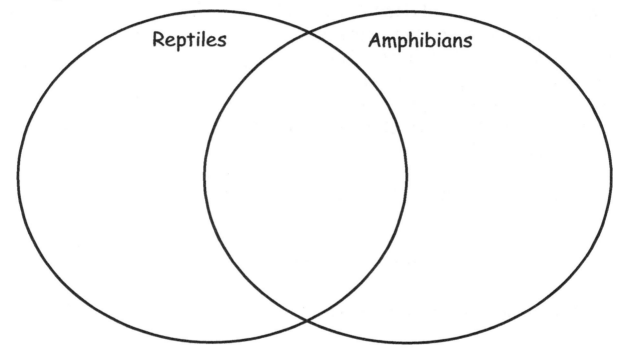

8. Generally, weather patterns in the United States move from _____ to _____.

9. Which clouds are thin and wispy and look like curls of hair?

    stratus   cirrus   cumulus   cumulonimbus

10. In which stage is the organism wrapped in a cocoon?

_____

Lesson #39

Soil Layers

Most soil has layers. Another word for these layers is **horizons**. The top layer (horizon) is called **topsoil**. Topsoil is made up of humus, mixed with sediment and mineral particles. **Humus** is the remains of decayed plants and animals. Plants grow best in this rich, dark layer. The next layer (horizon) of soil is called **subsoil**. Subsoil doesn't contain much humus. The soil particles are larger than those in topsoil. Subsoil contains clay and partially weathered rock. The final layer (horizon) of soil is called **bedrock**. Bedrock is beneath all of the other layers. This layer is solid rock.

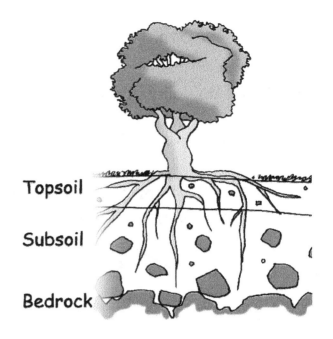

1. What is the name of the deepest soil horizon (layer)?

   _____

2. Name the three main horizons of soil.

   _____

   _____

   _____

3. _____ is the decayed remains of plants and animals.

   Subsoil   Topsoil   Humus   Horizon

Simple Solutions© Science Level 4

4 – 5. Label the diagram using the words below. Put each word next to an arrow.

water   sunlight   carbon dioxide   oxygen

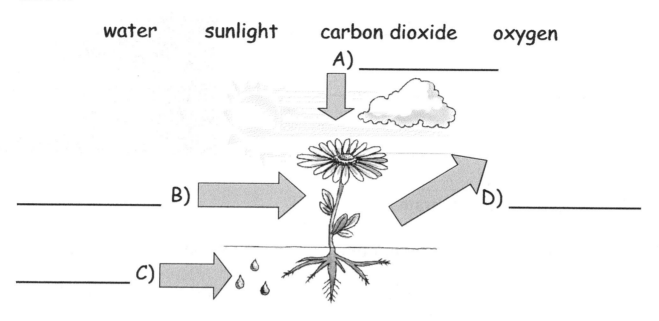

6. Which of these is not a basic need of animals?

    food   shelter   feet   water   air

7. An example of an amphibian is a _____.

    parrot   lizard   frog   snake

8. Nonvascular means _____.

    without soil   without plants   without tubes

9. A fox hunts and eats a rabbit. Here, the fox is the _____.

    predator   producer   prey

10. The term given to both the vent and the cone-like mountain left by the overflow of the erupted lava is _____.

    earthquake   volcano   horizon   landslide

Lesson #40

1. Name three forms of precipitation.

   _____

   _____

   _____

2. Which are invertebrates?

   sponge   chipmunk   worm   ant

   opossum   salamander   snail   giraffe

   scorpion   spider   salmon

3. Write T if the statement is true or F if it is false.

   _____ This is the formula for photosynthesis.
   carbon dioxide + water → sugar + oxygen

   _____ Spiders are insects.

4. Which instrument is used to measure wind speed?

   barometer   anemometer   thermometer

5. The remains of something that lived long ago is called a _____.

6. Which of these is a spinning column of air that touches the ground?

   blizzard   tornado   drought

Use this food chain to answer the questions below.

diatom → krill → arctic cod → seal → arctic wolf

7. Which animal is the final consumer in this food chain?

   _____

8. Which is the producer?   _____

9. What does the seal eat?   _____

10. Match each animal group to its definition.

    _____ Fish   A) moist skin; lives near water

    _____ Reptile   B) two legs; wings and feathers

    _____ Bird   C) lives whole life in water; breathes with gills

    _____ Amphibian   D) dry, scaly animal that lays eggs

    _____ Mammal   E) has hair or fur; feeds young with milk

Lesson #41

Types of Soil

Soil is a mixture of many different things. It is made up of bits of rock, minerals, and also bits of things that were once living. There are four different types of soil. They are sand, silt, clay, and loam. **Sand** has tiny grains of rock that you can easily see with your eye. **Silt** has grains of rock that are hard to see with your eye. **Clay** is a type of soil with the smallest soil particles. You might even need a microscope to see the grains in clay. **Loam** is a mixture of all of the types of soil: humus, clay, silt, and sand. Loam is the best soil to use when trying to grow vegetables and fruits. This is why most farms are built where the soil is loam.

1. List four types of soil.

    _____

    _____

    _____

    _____

2. Which type of soil is usually found on farms?

    silt   clay   sand   loam

Simple Solutions© Science   Level 4

3. What makes up humus?

   A) minerals   C) decayed plants and animals

   B) rocks   D) roots

4. Look at the picture to the right. What is shown here?

   a drawing   a fossil   a skull

5. Most earthquakes occur along _____.

   highways   mountains   faults   rivers

6. What type of weather brings heavy snow and strong, cold winds?

   hurricane   drought   thunderstorm   blizzard

7. The path of food from one living thing to another is a _____.

   nonvascular   food chain   photosynthesis

8. Jeremy wants to measure the wind direction. Which instrument should he use?

   anemometer   barometer   weather vane

9. The symbol for which type of front is shown?

   cold   warm   stationary

10. An (omnivore / herbivore) eats plants and other animals.

Lesson #42

Weathering

Weathering is the breakdown of rocks and minerals on the Earth's surface. Weathering breaks down rocks and minerals to smaller fragments or pieces. Water causes most weathering. The movement of water over the rock softens the rock, weakening it, which helps to wear the rock away. If you have ever gone to a stream or river and found smooth rocks (sediment) with rounded edges, these smooth, round rocks were caused by weathering. The running water which caused the rocks to scrape against each other made the sediment smooth and round.

Other things that can cause weathering are ice, temperature changes, and large, pounding ocean waves. Even plants can cause weathering. The roots of plants can grow in the cracks of rocks, splitting the rock as they grow.

1. The breakdown of rocks and minerals is called _____.

2. Name two things that can cause weathering.

   _____   _____

3. Which layer of soil is solid rock?

   bedrock   topsoil   subsoil   humus

4. Which is **not** a living thing?

   moss   pebble   root   mushroom

5. Match each type of front with its definition.

   _____ warm front   A) are shown on weather maps as a blue line with triangles on it

   _____ cold front   B) are shown on weather maps as a red line with half circles on it

   _____ stationary front C) fronts that stay in one place for many days

6. All plants are (producers / consumers) because they make their own food.

7. Which clouds are associated with powerful thunderstorms?

   cumulus   stratus   cumulonimbus   cirrus

8. Which of these has the greatest mass?

   a pop can   a desk   an eraser   a lady bug

9. How are roots and stems alike?

   A) Both move water through the plant.
   B) Both have chlorophyll.
   C) Both are green.

10. How are most mammals and birds alike?

    A) Both are cold-blooded.
    B) Both are warm-blooded.
    C) Both lay eggs.

Lesson #43

Erosion

After weathering breaks down rock, erosion happens. **Erosion** is the process by which the surface of the Earth gets worn down. It is the process of moving sediment. Erosion can be caused by moving water, ice, and wind. Rocks can be washed down a mountain, or rivers can wash small rocks (sediment) downstream. Large storm waves, along with the chemical make-up of the water, are what **erodes** (wears away) the rock along the coastline.

1. The process by which the surface of the Earth gets worn down is called _____.

2. A (cold front / warm front) usually moves slowly and brings steady rain, rather than thunderstorms.

3. List the five groups of vertebrates.

    _____   _____

    _____   _____

    _____

4. Name two ways seeds are spread.

    _____

    _____

5. The green substance inside leaves is called _____.

    photosynthesis   sugar   chlorophyll

6. Some animals go into a deep sleep for the winter. What is this called?

    camouflage   hibernation   migration

7. Most plants get their nutrients directly from _____.

    soil   air   sun   insects

8. What is a hypothesis?

    an experiment   a logical guess   an observation

Use the graph to answer the questions below.

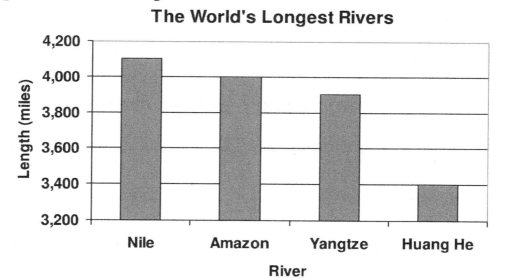

9. In the United States, the Missouri River is the longest river. It is about 2,500 miles long. A) How much longer is the Nile River than the Missouri River? B) How much longer than the Huang He River is the Nile?

    A) _____   B) _____

10. What is the total length of all four of the rivers in the graph? _____

Lesson #44

1. Which type of plants has tube-like structures that transport water from the roots to the stem and to the leaves?

    nonvascular   vascular   fungi

2. Which layer of soil lies below the topsoil?

    topsoil   bedrock   subsoil

3 – 4. Next to each vertebrate group, write a few words that describe that group. Two have been done for you.

Vertebrate	Description
amphibian	
bird	wings, 2 feet, and feathers
fish	
mammal	
reptile	dry, scaly skin, lay eggs, cold-blooded

5. Label the parts of a plant.

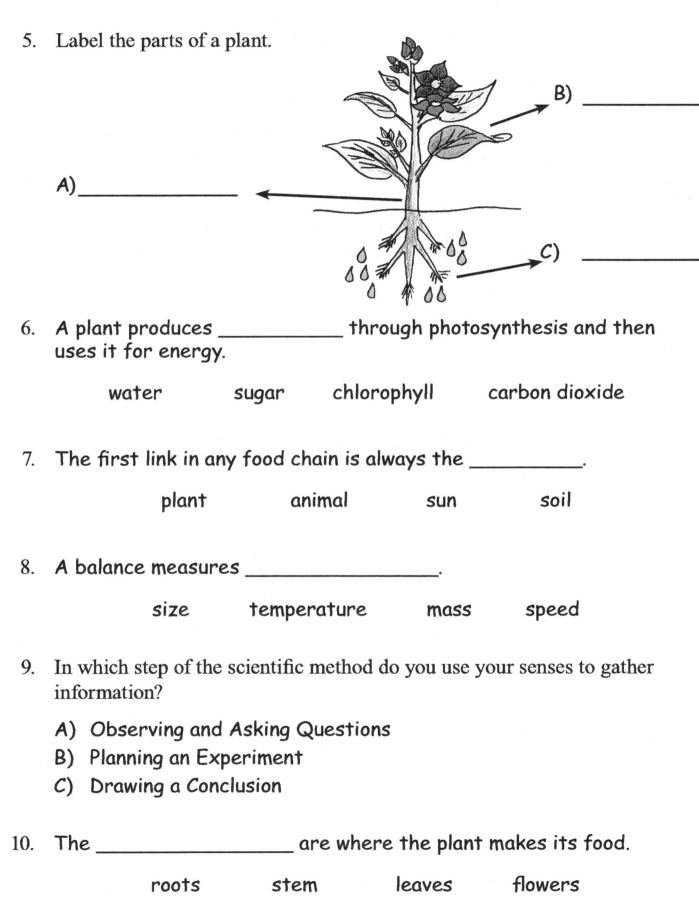

A) _____

B) _____

C) _____

6. A plant produces _____ through photosynthesis and then uses it for energy.

    water   sugar   chlorophyll   carbon dioxide

7. The first link in any food chain is always the _____.

    plant   animal   sun   soil

8. A balance measures _____.

    size   temperature   mass   speed

9. In which step of the scientific method do you use your senses to gather information?

    A) Observing and Asking Questions
    B) Planning an Experiment
    C) Drawing a Conclusion

10. The _____ are where the plant makes its food.

    roots   stem   leaves   flowers

Simple Solutions© Science   Level 4

Lesson #45

1. Which vertebrates have hair or fur and babies that are born alive?

    amphibians   reptiles   mammals   birds

2 – 3. For each organism in the chart below, decide whether it is a **producer** or a **consumer**, and put a ✓ in the correct column.

	Producer	Consumer
grass		
algae		
owl		
bear		
seaweed		

4. A large body of ice that moves slowly down a slope is called a ____.

    volcano   glacier   earthquake   hail storm

5. Choose two examples of fungi.

    mushroom   moss   dandelion   ivy   mold

6. A person who studies earthquakes is called a _____.

    quakeologist   Richterologist   seismologist

Simple Solutions® Science — Level 4

Use the *Fact Cards* in the Help Pages to fill in the chart. Then answer the questions below.

Invertebrate	Diet	Number of Eggs
grasshopper		
cricket		
housefly		

7. Which invertebrates eat leaves? _____

8. How many eggs does a housefly lay? _____

9. Name two things that a housefly eats.

   _____   _____

10. Which of these is **not** a form of precipitation?

    A) rain   C) pupa

    B) sleet   D) snow

Lesson #46

1. This diagram shows the stages of complete _____.

    A) photosynthesis   C) migration

    B) hibernation   D) metamorphosis

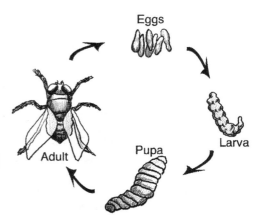

2. Which type of cloud is shown?

    A) cirrus   C) cumulonimbus

    B) cumulus   D) stratus

3. Plants grow best in which layer of soil?

    subsoil   topsoil   bedrock

Some harmless animals have to imitate other animals to stay alive. These "copycats" are called **mimics**. Imitating the look of another animal is called **mimicry**. If insects or other less powerful animals can trick their predators into thinking they are a different animal by imitating another animal, they just might survive a little longer.

4. What is it called when an animal imitates the look of another animal?

    _____

5. Which type of front does the symbol show?

    _____

6 – 7. Use the Venn diagram below to compare and contrast fish and reptiles. List two similarities and one difference.

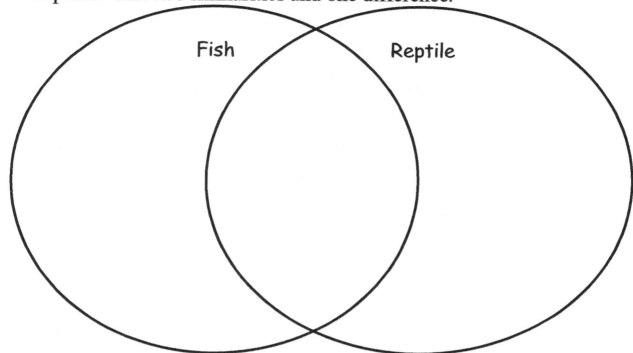

8. A student thinks that a piece of clay will float in a tub of water. This is an example of which step of the scientific method?

   A) Forming a Hypothesis
   B) Conducting an Experiment
   C) Drawing a Conclusion

9 – 10. Use the word bank to complete the sentences below.

| oxygen | water | photosynthesis |
| sunlight | sugar | carbon dioxide |

During the process of _____, plants take in

_____, _____, and _____.

Plants make _____ and give off _____ as a by-product of photosynthesis.

Simple Solutions© Science Level 4

Lesson #47

1. Match each tool with its description.

   ____ balance   A) magnifies, or makes something larger

   ____ anemometer   B) measures the mass of an object

   ____ hand lens   C) measures how hot or cold something is

   ____ thermometer   D) measures wind speed

Use this food chain to answer the questions below.

polar bear ← Arctic cod ← krill ← diatom

2. Which is the final consumer? _____

3. What is the producer in this food chain? _____

4. **A tremor or shaking of the earth's surface usually caused by movement of rock in the crust is a(n) _____.**

   volcano   tornado   earthquake   landslide

5. Which of these would an herbivore eat?

   worm   fish   seeds   mouse

94

6. The process by which the surface of the Earth gets worn down is called _____.

        evaporation   photosynthesis   erosion

7 – 8. Read the description of behaviors that help animals survive. Use the words below to fill in the chart.

    hibernate   migrate   camouflage   instinct

Description	Trait
a behavior an animal knows without being taught	
to go into an inactive state during the winter	
to travel from one place to another and back again	
anything that disguises an animal and helps it hide	

9. Which type of vertebrate begins its life breathing with gills and eventually breathes with lungs?

        reptile   amphibian   mammal   fish

10. What is the green substance inside leaves called?

        chlorophyll   sugar   photosynthesis   vascular

Lesson #48

Water Cycle

Every organism needs water, and water is continually moving through the environment. Water recycles itself in a process called **the water cycle**. The diagram shows that the water cycle has no beginning or end. This is how the water cycle works:

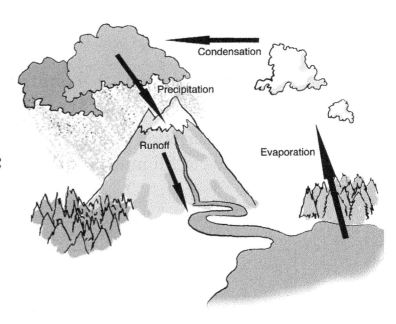

Water on the Earth's surface **evaporates**, becoming water vapor. The atmosphere is filled with **water vapor**. As water vapor cools, it **condenses**, forming clouds. When enough water condenses, heavy rain clouds give way to **precipitation** that falls to the Earth's surface. Precipitation can come in the form of rain, snow, sleet, or hail. Water from precipitation or melting ice and snow sinks into the ground or runs into oceans, lakes, rivers, and other bodies of water. This continues the water cycle.

1. The movement of water through the environment is called

   the _____.

2. In the water cycle, what happens before water condenses?

   A) Water vapor becomes a gas.
   B) Water evaporates.
   C) Precipitation falls to Earth.

3. Write T if the statement is true or F if it is false.

   _____ The water cycle has no end.

4. What is pictured here?

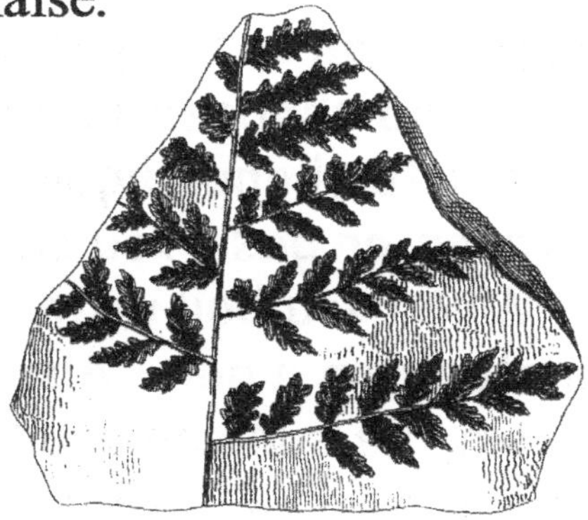

   _____

5. The symbol for what kind of front is shown?

   _____

6. What layer of soil is pure rock?

          subsoil   topsoil   bedrock

7. Which step of the scientific method comes right after **Forming a Hypothesis**?

   _____

8. Which of these is **not** a type of soil?

       loam   sand   rock   silt   clay

9. What do you call an organism that eats other living things in order to get energy?

       consumer   producer   metamorphosis

10. Which stage of metamorphosis is shown?

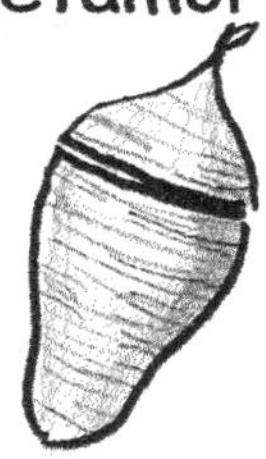

    _____

Simple Solutions© Science   Level 4

Lesson #49

Parts of the Water Cycle/Evaporation

Evaporation is the process in which a liquid changes to a gas. It takes heat to change liquid water into a vapor, or gas. You cannot see a gas. When you boil water, it can evaporate from a pan. Also, the heat from the sun can cause water to evaporate from oceans, lakes, and rivers. After a rainstorm, you may see puddles of water. Once the sun comes out, after a short while the water puddles dry up, or evaporate, and the water is now an invisible gas – water vapor. The water vapor can travel high up into the air.

1. The process where a liquid changes to a gas is called _____.

2. You notice clouds forming on a warm, sunny day. What can you infer is happening in the atmosphere?

   _____

   What might happen later in the day?

   _____

3. The movement of water through the environment is called the _____.

    condensation   evaporation   water cycle

Simple Solutions© Science   Level 4

4. To which group does this animal belong?

   A) reptile   C) mammal

   B) amphibian   D) invertebrate

   scorpion

5. Which is **not** a basic need of animals?

   food   antennae   water   shelter

6. What is the correct order for complete metamorphosis?

   A) larva ➡ egg ➡ pupa ➡ adult
   B) pupa ➡ adult ➡ egg ➡ larva
   C) egg ➡ larva ➡ pupa ➡ adult
   D) egg ➡ pupa ➡ larva ➡ adult

7. Look at the balance. Which has more mass, the sparrow or the hummingbirds?

   _____

8. Which of these is a by-product of photosynthesis?

   helium   carbon dioxide   oxygen   chlorphyll

9. An animal that hunts another animal for food is called a(n) _____.

   prey   predator   herbivore   producer

10. Animals with a backbone are called _____.

Lesson #50

Parts of the Water Cycle/Condensation

In the last lesson, we ended by saying that water vapor can go high up into the air. As water vapor moves higher into the air, it becomes cooler. When the water vapor cools enough, condensation occurs. **Condensation** is the process where water vapor turns into liquid water. The water vapor changes to tiny droplets of liquid water. **Dew** on the morning grass is a type of condensation. There may be condensation on the bathroom mirror after taking a shower, or on the outside of a cold bottle (as seen in the picture). **Clouds** and **fog** are other examples of condensation.

Condensation

1. The process where water vapor turns into liquid water is called _____.

2. Name two examples of condensation.

   _____

   _____

3. Which ecosystem will probably have the **least** plant life?

       forest   grassland   desert   freshwater

4. A break in the crust where rock moves is called a _____.

       volcano   shield   fault   fossil

5. What does vascular mean?

       with a vase   with tubes   with cones   with flowers

Look at the weather map and use it to answer the questions below.

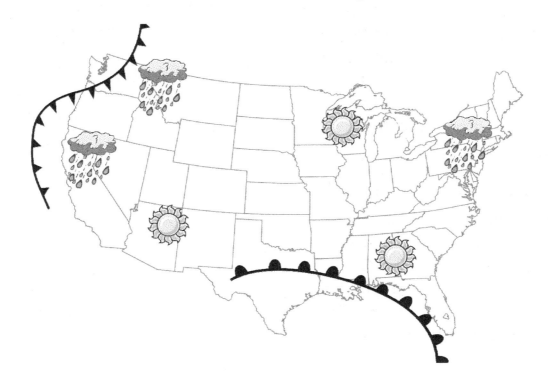

6. Which part of the country is experiencing a warm front?

   the northwest   the Midwest   the southeast

7. According to the map, which of these states is having sunny weather?

   California   Pennsylvania   Georgia   New York

8. To which group do crabs and lobsters belong?

   A) reptiles   C) invertebrates
   B) fish   D) vertebrates

9 – 10. Name the four basic needs of plants.

   _____   _____

   _____   _____

Lesson #51

Parts of the Water Cycle/Precipitation

When water vapor condenses, it forms clouds. Inside the clouds, these tiny droplets of water can join together to form bigger droplets of water. The bigger droplets get even bigger and heavier. When these droplets become too heavy to remain in the air, they fall to Earth as precipitation. **Precipitation** is any form of water that falls to the ground. The kind of precipitation depends on the temperature of the air. Some of the water from precipitation is used by plants and some flows to the bodies of water on Earth's surface such as lakes, rivers, and oceans.

1. Any form of water that falls to the Earth is called _____.

2. What makes up humus?

   A) decayed plants and animals
   B) rocks
   C) minerals
   D) roots

3. Choose the vertebrates.

   anteater   chipmunk   rat   ant

   opossum   crab   snail   giraffe

4. Which word means *an educated guess*?

   observation   hypothesis   experiment

Simple Solutions© Science Level 4

5 – 7. Use the *Fact Cards* in the Help Pages to complete the Venn diagram comparing and contrasting a katydid and a cockroach. Give two similarities and two differences.

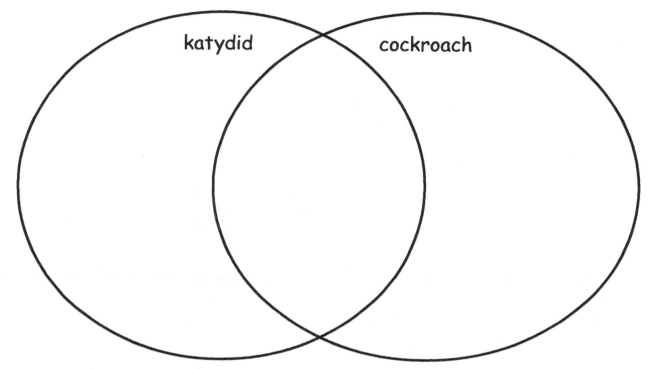

8. Some examples of fungi are _____.

    A) molds and mushrooms
    B) flowering plants and trees
    C) ferns and mosses

9. Draw the symbol for warm front in the box.

10. What are two traits of mammals?

    _____

    _____

Lesson #52

1. The process where a liquid changes to a gas is called _____.

   precipitation   evaporation   condensation

2. Leon wants to measure the mass of an object. Which instrument should he use?

   anemometer   balance   barometer   thermometer

3. Which clouds are puffy clouds that sometimes look like pieces of floating cotton?

   stratus   cirrus   cumulonimbus   cumulus

Hurricanes are large, tropical storms with wind speeds of 74 miles an hour or more. Hurricanes form over warm water in the tropical oceans. Hurricanes can last for weeks when they remain at sea. Once a hurricane reaches land and it's not getting the energy from the warm water, it tends to weaken. The center of the hurricane is called the "eye." Like tornadoes, hurricanes are categorized by their wind speed. Category 1 is the weakest in strength and Category 5 is the strongest, with winds at 155 mph or greater.

4. What is the center of a hurricane called? _____

5. Where do hurricanes form?

_____

6. Which are the two main groups of plants?

    A) vascular and nonvascular
    B) vertebrates and invertebrates
    C) ferns and fungus

7. Put a ✓ next to each statement that is true about the water cycle.

    A) _____ Water recycles itself.

    B) _____ The water cycle has a definite end.

    C) _____ Water on the Earth's surface evaporates, becoming water vapor.

8. What three things are needed for photosynthesis to occur?

    A) carbon dioxide, sugar, and sunlight
    B) sunlight, water, and oxygen
    C) carbon dioxide, water, and sunlight

9. Vascular plants are divided into flowering, cone-bearing, and _____.

    mushrooms   mosses   ferns

10. A large body of ice that moves slowly down a slope is called a(n) _____.

    A) erosion
    B) landslide
    C) glacier

Lesson #53

1. Match each cloud type with its description.

   _____ cirrus   A) uniform grayish clouds that often cover the entire sky

   _____ cumulus   B) thin, wispy clouds; they look like wisps of hair

   _____ stratus   C) associated with powerful thunderstorms

   _____ cumulonimbus   D) puffy clouds that sometimes look like pieces of floating cotton

2. Which of these has the greatest mass?

   a pencil   a toothpick   a wiffle ball   a baseball bat

3. Which layer of soil is best for growing plants?

   bedrock   topsoil   subsoil

4. Arrange the steps of the scientific method in the correct order.

   A) Conducting an Experiment   1. _____

   B) Drawing Conclusions   2. _____

   C) Planning an Experiment   3. _____

   D) Forming a Hypothesis   4. _____

   E) Observing and Asking Questions   5. _____

5. Which list correctly shows how energy moves in a food chain?

   A) flower ⟶ deer ⟶ human ⟶ sun

   B) deer ⟶ flower ⟶ sun ⟶ human

   C) human ⟶ deer ⟶ flower ⟶ sun

   D) sun ⟶ flower ⟶ deer ⟶ human

6. What is it called when an animal imitates the look of another animal?

      instinct   migration   mimicry   hibernation

7. Plants make their own food through a process called _____.

      erosion   photosynthesis   humus   weathering

8. After the sun, all food chains begin with a _____.

      consumer   producer   omnivore

9. Which type of plant has tube-like structures that transport water from the roots to the stem and to the leaves?

      vascular   nonvascular   fungi

10. Put a ✓ next to each statement that is true about the food chain.

    _____ All animals depend on plants for their energy.

    _____ Animals are producers.

    _____ The sun is always the first link in any food chain.

Lesson #54

Ecosystem

All living and nonliving things in an area form an **ecosystem**. You may have learned last year that living things need food, water, and air to survive. Some examples of living things in your environment are plants, animals, and humans. The nonliving things in your environment include things like weather, soil, air, and water.

Ecosystems can be very small, like the space under a log where you might find soil and tiny insects or bacteria. An ecosystem can also be as large as a forest, where there is an abundance of plant and animal life.

1. The living and nonliving things in an area make up an _____.

2. What three things do all living things need to survive?

   _____

   _____

   _____

3. What is the center of a hurricane called?

       funnel   precipitation   eye   spiral

4. To which group does this animal belong?

   A) reptile   C) mammal

   B) amphibian   D) fish

5. The animal pictured above is (warm / cold) blooded.

6. Write T if the statement is true or F if it is false.

   _____ Insects are the smallest group of invertebrates.

7. What can happen when an area doesn't get enough rain?

   flood   drought   landslide   hurricane

8. Which of these animals is an herbivore?

   tiger   wolf   sheep   hawk

9. Which step of the scientific method comes right after **Conducting an Experiment**?

   _____

10. Match each type of front with its definition.

    _____ stationary front   A) front that stays in one place for many days

    _____ warm front   B) shown on weather maps as a red line with half circles on it

    _____ cold front   C) shown on weather maps as a blue line with triangles on it

Lesson #55

Population

One oak tree is an individual. One sunflower is an individual. You are also an individual. A **population** is a group made up of the same kind of individuals living in an ecosystem. A group of oak trees is a population. A cluster of sunflowers is a population and a group of people living in the same town is a population. Oak trees live in the same forest as evergreen trees. Oak trees are a different kind of tree than evergreens, so they belong to a different population.

1. A group made up of the same kind of individuals living in an ecosystem is a _____.

2. The living and nonliving things in an area make up a(n) _____.

    ecosystem   organism   water cycle

3. What do most reptiles, amphibians, and fish have in common?

    A) They have feathers.
    B) They don't have a backbone.
    C) They lay eggs.
    D) They have scales.

4. The green substance inside leaves is called _____.

    chlorophyll   photosynthesis   sugar

5. What do these animals in the picture have in common?

    A) They are all amphibians.
    B) They are all cold-blooded.
    C) They are all invertebrates.

6. Where do hurricanes get their energy?

    from the wind   from the sun   from the warm water

7. An instrument that shows the movement of the Earth's surface during an earthquake is called a _____.

    anemometer   seismograph   barometer

8. What is a hypothesis?

    an experiment   a logical guess   an observation

9 – 10. Use the graphic organizer to fill in four characteristics of reptiles.

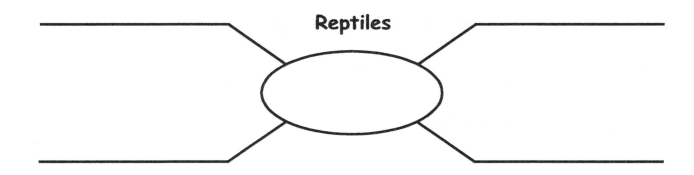

Simple Solutions© Science   Level 4

Lesson #56

Community

A **community** is made up of all the populations that live in the same place. Many populations make up the communities in the desert ecosystem. The plants of the desert include Saguaro trees and cacti. Desert animals include rattlesnakes, roadrunners, Gila monsters, camels, beetles, and bees. The plants and animals in a community depend on each other for food, shelter, and the spreading of seeds.

1. All of the populations living in the same place make up a _____.

2. The _____ are where the plant makes its food.

    stem   roots   flowers   leaves

3. The series of changes in appearance that some organisms go through is called _____.

    metamorphosis   vertebrates   vascular

4. Nonvascular means *without* _____.

    seeds   *tubes*   *flowers*   *stems*

5. Name two forms of precipitation.

    _____   _____

6 – 7. Next to each description, write **insect** or **spider**.

Description	Insect or spider?
has 3 body parts	
has 8 legs	
has 2 body parts	
has 6 legs	

8. In this food chain, which is the producer? _____

   tulip ➡ ant ➡ spider ➡ robin ➡ cat

   Which is the final consumer? _____

9. An animal without a backbone is called a(n) _____.

   vertebrate   mammal   invertebrate

10. Which of the following is a trait of mammals?

    A) feathers   C) gills
    B) fur   D) scales

Lesson #57

1. Which of the following causes erosion?

   A) moving water

   B) evaporation

   C) humus

2. Besides silt and loam, list the two other types of soil.

   A) s_____

   B) c_____

3. Generally, weather patterns in the United States move

   from _____ to _____.

4. A plant produces _____ through photosynthesis and then uses it for energy.

   oxygen   sugar   water   carbon dioxide

5. Brenda enjoys fresh vegetables like peppers, carrots, and beans. She also eats strawberries, apples, and peaches. Brenda enjoys eating hamburgers, chicken, and steak. Which type of consumer is Brenda?

   herbivore   carnivore   omnivore   none of these

6. What is the process in which water vapor turns into liquid water?

	condensation		evaporation		precipitation

7. What is a large body of air called?

	wind		storm		air mass		stationary front

8. Match each animal group to its definition.

	____ Bird		A) has hair or fur; feeds young with milk

	____ Reptile		B) two legs, wings, and feathers

	____ Mammal		C) moist-skin, lives near water

	____ Amphibian		D) dry, scaly animal that lays eggs

	____ Fish		E) lives whole life in water, breathes with gills

9. To which group do insects and sponges belong?

	A) reptiles		C) invertebrates
	B) fish			D) vertebrates

10. During the first step of the scientific method, what do you use to gather information?

	a microscope		your senses		the internet

Lesson #58

Rainforest

Tropical **rainforests** are found in a band along the equator. They are dense forests that get high amounts of rain each year. There are rainforests in South and Central America, Africa, the islands around Australia, and in Asia. Rainforests are wet and warm. They are home to millions of plants and animals. A few of the animals found in the rainforest include macaw, toucan, poison arrow frog, leaf cutter ant, spider monkey, and piranha. However, there are more insects than any other animal in the rainforest.

Some foods that come from the rainforest include cashews, bananas, pineapple, coffee, tea, yams, cinnamon, cocoa, and peanuts. The rainforest gets 7 to 33 feet of rain each year. That's a lot of rain!

1. Name two places where rainforests can be found.

   _____

   _____

2. What is the most numerous animal of the rainforest? _____

3 – 4. Name three products that come from the rainforest.

   _____   _____   _____

5. All of the populations living in the same place make up a _____.

   population   ecosystem   community

6. How can the roots of plants cause weathering of rocks?

   A) It holds the rock in place.
   B) It cracks them.
   C) It makes them softer.

7. This rock shows fossils of water plants and shells. What does this tell us about the rock?

   A) This rock is dirty.
   B) This rock is softer than most rocks.
   C) This rock was once at the bottom of the sea.
   D) This rock is stronger than other rocks.

8. Kendra has a soil sample containing living and nonliving materials. Which material was once living?

   water droplets   sand   decomposing leaves   pebbles

9. In order to survive, all animals need _____.

   A) feet, wings, and eyes
   B) light, nutrients, and soil
   C) food, water, and air

10. All animals are (producers / consumers) because they get their food from plants.

Lesson #59

Tundra

The tundra is cold and treeless. It is the coldest ecosystem. It gets very little rain or snow. The word **tundra** means *treeless plain*.

Arctic tundras are located in Greenland, Alaska, Canada, Europe, and Russia. **Alpine tundras** are located high in the mountains around the world. Some animals found in the arctic tundra include caribou, lemmings, polar bears, wolves, snowy owls, musk oxen, and Arctic foxes. Arctic bumblebees, flies, grasshoppers, and mosquitoes are some insects found in the tundra.

The ground in the tundra is covered by a layer of frozen subsoil called **permafrost**. Permafrost is ground that has remained at or below 32°F or 0°C for two or more years in a row. Some low shrubs, reindeer mosses, liverworts, grasses, and about 400 varieties of flowers can survive in the tundra despite the cold temperatures and frozen soil.

1. What does the word *tundra* mean?

   _____

2. What is the frozen layer of subsoil called?

   _____

3. Name two animals found in the tundra.

   _____

   _____

4. What do you call dense forests that get a lot of rain each year?

    desert   grassland   rainforest   tundra

5. Match each part of the water cycle with its description.

    ____ evaporation   A) water falling to Earth's surface

    ____ precipitation   B) water vapor forming droplets

    ____ condensation   C) water turning to vapor

6. When two different large air masses meet, a _____ is formed.

    cloud   hurricane   front   fault

7. Choose the correct formula for photosynthesis.

    A) carbon dioxide + nutrients + water → sugar

    B) oxygen + water → sugar + carbon dioxide

    C) carbon dioxide + water → sugar + oxygen

8. Write **T** if the statement is true or **F** if it is false.

    _____ Nonvascular plants do not have tubes to carry water and food to the parts of the plant.

9. The symbol for which type of front is shown?

    _____

10. A balance measures the (air pressure / mass) of an object.

Simple Solutions© Science   Level 4

Lesson #60

Taiga

The **taiga** is a cold forest. Taiga is a Russian word that means *forest*. The taiga can be found in parts of North America, Europe, and Asia. You will usually find taigas south of the tundra and north of the deciduous forest. The taiga is the largest ecosystem in the world. There are not many animals in the taiga because of the harsh climate. Some animals that can be found in the taiga include the lemming, Arctic fox, Canada goose, and reindeer.

1. What is the largest ecosystem in the world? _____

2. What does the word *taiga* mean? _____

3. What is the role of a producer in a food chain?

    A) eat plants
    B) prey on other animals
    C) make food

4. What is shown in the diagram to the right?

    A) a population
    B) a community
    C) a life cycle
    D) a food chain

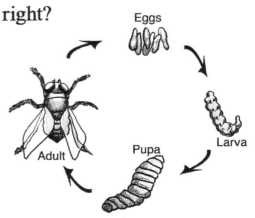

Use the water cycle diagram to answer the questions below.

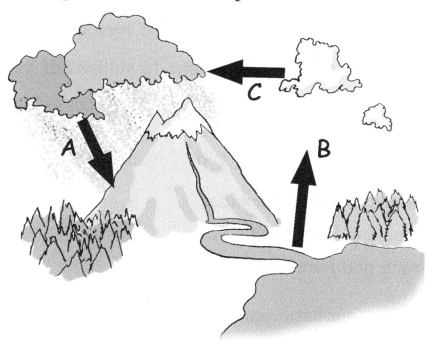

5. What process is occurring at **B**?

   precipitation   evaporation   condensation

6. What process is occurring at **A**?

   precipitation   evaporation   condensation

7. All of the populations living in the same place make up a _____.

   ecosystem   population   community

8. Write T if the statement is true or F if it is false.

   _____ Sunlight, water, and carbon dioxide are necessary for photosynthesis to occur.

9. The first link in any food chain is always the _____.

10. An (herbivore / omnivore) eats plants and other animals.

Lesson #61

1. Match each ecosystem with its definition.

   ____ taiga   A) cold and treeless

   ____ rainforest   B) home to millions of plants and animals

   ____ tundra   C) the largest ecosystem in the world

2. Which two storms are categorized by wind speed?

   A) thunderstorm and tornado
   B) tornado and hurricane
   C) blizzard and hurricane

3. Animals have four basic needs. Two are given. Name the others.

   air

   water

   _____

   _____

4. What do you call the part of the plant that grows underground and takes water and nutrients from the soil?

   stem   leaf   root   seed

5. Seeds can be spread by wind, by water, and by _____.

   gills   animals   metamorphosis

Simple Solutions© Science   Level 4

Every time we make or build something, we are using some of the Earth's resources. In order to help save some of these resources, try to reuse or recycle things. The symbol to the right means that a product can be **recycled**. When you recycle, you are helping to conserve resources. When an item is recycled, it is broken down or changed in some way so that the material can be used to make something new. Some materials that can be recycled are **paper, aluminum cans**, some **plastics**, and **glass**.

6. Draw the symbol for recycling in the box.

7 – 8. List four things that can be recycled.

_____   _____

_____   _____

9. A group made up of the same kind of individuals living in the same ecosystem is a _____.

        community   ecosystem   population

10. An example of a reptile is a _____.

        robin   salmon   ostrich   snake

Lesson #62

1. The frozen layer of subsoil is called _____.

    taiga   permafrost   sleet   hail

2. A robin eats a worm. In this situation, the worm is the _____.

    predator   producer   prey

3. Which animal is an invertebrate?

    pig   snail   monkey   shark

4. What gas do plants give off during photosynthesis?

    carbon dioxide   sugar   water   oxygen

5. Label the parts of a plant.

    _____

    _____

    _____

6. Which are the two main groups of plants?

    A) vascular and nonvascular

    B) vertebrates and invertebrates

    C) ferns and fungi

7. The chart shows two categories. Put each organism in the list in the correct category.

   algae   fox   worm   dandelion   moss   fish

Producer	Consumer

8. What does this tool measure?

   A) air pressure
   B) temperature
   C) rainfall

9. Which step of the scientific method comes right after **Observing and Asking Questions**?

   _____

10. Which of these is **not** a type of soil?

    sand   loam   silt   pebble   clay

Lesson #63

1. Which organism would most likely be the final consumer in this food chain?

    worm   fox   robin   sunflower

2. The process in which a liquid changes to a gas is called _____.

    precipitation   evaporation   condensation

3. Which clouds are associated with powerful thunderstorms?

    cumulus   cumulonimbus   stratus   cirrus

4. What does this symbol mean?

    _____

5. Put a ✓ next to each item that can be recycled.

glass bottle	
T-shirt	
newspaper	
Styrofoam cup	
phone books	

6. What do spiders and insects have in common?

   number of legs   outer body covering   wings

7. Anthony took a survey. He is making a graph to show the favorite flavor of juice among his friends. The graph is an example of which part of the scientific method?

   A) Observing and Asking Questions
   B) Planning an Experiment
   C) Forming a Hypothesis
   D) Drawing Conclusions

8. All of the populations living in the same place make up a(n) _____.

   population   community   ecosystem

9. To which group does this animal belong?

   A) reptile
   B) amphibian
   C) mammal
   D) invertebrate

10. Name the five vertebrate groups.

    _____

    _____

    _____

    _____

    _____

Lesson #64

1. In which stage of metamorphosis is this organism?

    larva   egg   pupa

2. Fill in the missing steps of the scientific method.

    1. _____

    2. _Forming a Hypothesis_____

    3. _____

    4. _Conducting an Experiment____

    5. _____

3. The green substance inside leaves is called _____.

    chlorophyll   sugar   loam   silt

4. _____ is the breakdown of rocks and minerals.

    Photosynthesis   Weathering   Condensation

5. Animals have four basic needs. List them.

    _____ _____

    _____ _____

6. The movement of water through the environment is called the _____.

        precipitation   evaporation   water cycle

7. Put a ✓ next to each statement that is true.

    _____ Insects are the most numerous animal in the rainforest.

    _____ The tundra is the largest ecosystem in the world.

    _____ The ground in the tundra is covered with permafrost.

8. Which has more mass, the parrot or the sparrow?

_____

9. Draw the symbol for a warm front in the box.

10. What is the correct order for complete metamorphosis?

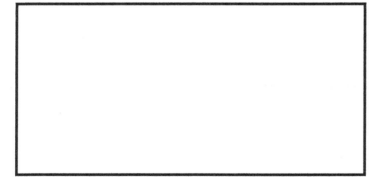

    A) larva → egg → pupa → adult
    B) egg → larva → pupa → adult
    C) pupa → adult → egg → larva
    D) egg → pupa → larva → adult

Lesson #65

States of Matter: Solid

Matter is anything that takes up space. Matter is all around you. The things you can see and even the things you can't see are matter. Fog, ice, and rain are all examples of matter. Matter is found in different states, or forms. The three states of matter are **solid**, **liquid**, and **gas**. Matter can change from one state to another by heating or cooling.

A **solid** is matter that has a definite shape and takes up a definite amount of space. A table, desk, computer, couch, and even your brothers or sisters are matter.

1. Anything that takes up space is called _____.

2. Name the three states of matter.

   _____   _____   _____

3. Name two examples of matter in the picture above.

   _____

   _____

4. What word means *the colors and patterns an animal uses to disguise itself*?

         instinct   camouflage   migrate

5 – 7. Look at the chart below. Read the description of each ecosystem. Write the name of the ecosystem that matches its description.

desert   tundra   rainforest

taiga   grassland   deciduous forest

Description	Ecosystem
the trees lose their leaves in the fall	
largest ecosystem in the world	
less than 10 inches of rain each year; limited plant life	
dry and usually flat; food crops grow well here	
cold and treeless; permafrost found here	
abundant plant and animal life; located near the equator	

8. A plant produces _____ through photosynthesis and then uses it for energy.

   water   oxygen   sugar   carbon dioxide

9. What is shown here?

   _____

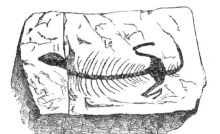

10. A (mushroom / fern) is an example of a fungus.

Simple Solutions® Science Level 4

Lesson #66

States of Matter: Liquid & Gas

A **liquid** takes the shape of its container. You can take a glass of water and pour it into a long, thin vase. The amount of water hasn't changed, but the shape has changed. The water takes on the shape of the vase. Shampoo, juice, paint, and glue are all examples of liquids. A **gas** does not have a definite shape or take up a definite amount of space. A gas will spread out to fill its container. Oxygen, helium, and carbon dioxide are examples of gases.

1. This type of matter has a definite shape and takes up a definite amount of space. What is it?

        solid   liquid   gas

2. Write **T** if the statement is true or **F** if it is false.

    _____   A liquid takes the shape of its container.

3. Give two examples of a liquid.

    _____

    _____

4. Which is a living thing?

        cloud   water   root   carbon dioxide

5. Which is the second step of the scientific method?

   A) Observing and Asking Questions
   B) Forming a Hypothesis

6. Which instrument should you use to measure wind speed?

   anemometer   balance   barometer   thermometer

7. Match each stage of a butterfly metamorphosis with its definition.

   _____ larva   A) the caterpillar is wrapped in a cocoon

   _____ pupa   B) the life cycle begins here

   _____ egg   C) the egg hatches into a caterpillar

   _____ adult   D) a butterfly comes out of the cocoon

8. Which organism would most likely be the final consumer in this food chain?

   frog   caterpillar   dandelion   snake

9. In item 8, which would be the producer? _____

10. Which animal would best fit into this ecosystem?

    camel   elephant   frog   penguin

Lesson #67

Physical Change

A physical change is a change that doesn't create something new. When you cut or shred paper, these are physical changes. You may have changed the size or shape of the paper, but it is still paper. Likewise, ice, water, and steam are all water, but in different forms. If you cut a log with a saw, it is still a log, only shorter. You have not changed the substance.

1. A change that doesn't create something new is a _____ change.

2. Name the three states of matter.

   _____

   _____

   _____

3. Write T if the statement is true or F if it is false.

   _____ Matter is anything that takes up space.

   _____ The two main kinds of plants are vascular and flowering.

4. During photosynthesis, a gas is given off. Which gas is it?

    helium   oxygen   carbon dioxide

5. What do you call a large body of air?

    wind   thunderstorm   cold front   air mass

6. A large body of ice that moves slowly down a slope is called a _____.

    landslide   mudslide   glacier   erosion

7. What is the role of a producer in a food chain?

    A) eat other animals   C) eat plants

    B) eat decayed animals   D) make food

8. Which ecosystem gets 7 – 33 feet of rain per year?

    tundra   taiga   grasslands   rainforest   desert

9. In a swamp, many different organisms live together. A swamp is an example of a(n) _____.

    A) population   C) plant system

    B) ecosystem   D) community

10. What does a carnivore eat?

    plants   sticks   rocks   animals

Lesson #68

1. Two vertebrate groups are warm-blooded. Which are they?

    birds   fish   reptiles   mammals   amphibians

2. What process is shown?

    condensation   evaporation   precipitation

3. When amphibians are born, they are most like _____.

    birds   mammals   fish

4. Look at the words below. Put each word under the correct heading.

    snow   mushroom   oxygen   juice   octopus

Living	Nonliving

5. What three things are needed for photosynthesis to occur?

    A) carbon dioxide, water, and sunlight

    B) carbon dioxide, sugar, and sunlight

    C) sunlight, sugar, and oxygen

6. Which layer of soil lies below the topsoil?

        subsoil   bedrock   pebbles

7. To which group do ants and beetles belong?

    A) reptiles   C) insects

    B) vertebrates   D) spiders

8. Which clouds are thin and wispy and look like curls of hair?

      stratus   cumulus   cumulonimbus   cirrus

9 – 10. Put the steps of the scientific method in the correct order.

    A)   Conducting an Experiment   1) _____

    B)   Forming a Hypothesis   2) _____

    C)   Planning an Experiment   3) _____

    D)   Drawing Conclusions   4) _____

    E)   Observing and Asking Questions   5) _____

Lesson #69

Chemical Change

In a **chemical change** the make-up of a substance is changed into one or more new substances. Some examples of chemical changes include burning a log, cooking an egg, baking a cake, or the rusting of a metal chair. Another example is when a half-eaten apple turns brown. In the example of a log burning (top), as the wood burns, it turns into a pile of ashes and gases. After this wood is burned, (bottom) you cannot put it back into its original form as a log. The way to tell whether something is a physical change or a chemical change is that in a physical change the composition of a substance does not change and in a chemical change the composition of a substance does change.

1. When substances are changed into different substances, this is an example of a _____ change.

2. List two examples of a chemical change.

   _____

   _____

3. Write T if the statement is true or F if it is false.

   _____ In a physical change, the make up of the substance does not change.

4. Which is not a trait of adult mammals?

   A) give birth to their young alive
   B) fur
   C) gills
   D) feed milk to their young

5. Draw the symbol that means *recycle*.

6. Which invertebrate has 8 legs and 2 body parts?

   _____

7. When two different large air masses meet, a _____ is formed.

   fault   front   hurricane   cloud

8. Which ones would most likely be found in the rainforest?

   toucan   polar bear   poison arrow frog   camel   caribou

9. A group of oak trees is an example of a(n) _____.

   community   population   ecosystem

10. What is the process in which water vapor turns into liquid water?

    A) condensation
    B) precipitation
    C) evaporation

Lesson #70

1 – 2. Decide whether the phrases below describe a physical change or a chemical change, and then put each under the correct heading in the chart.

Remember: In a physical change the composition of a substance does not change, but in a chemical change the composition of a substance does change.

sawing a log   shredding paper   burning a log

melting ice   baking a cake   a rusting can

	Physical Change	Chemical Change
A)		
B)		
C)		
D)		
E)		
F)		

3. An anemometer measures the (air pressure / wind speed).

4. All living things have three basic needs. What are they?

_____ _____

_____

5 – 6. Read the description of traits that help animals survive. Use the words below to fill in the chart.

mimicry   hibernate   migrate   camouflage   instinct

Description	Trait
imitating the look of another animal	
to travel from one place to another and back again	
to go into an inactive state during the winter	
anything that disguises an animal and helps it hide	

7. What helps us to learn about plants and animals of the past?

   weather   fossils   footprints   clouds

8. An organism without a backbone is called a(n) _____.

   invertebrate   omnivore   vertebrate

9. Vertebrates that are cold-blooded with dry, scaly skin are _____.

   amphibians   mammals   fish   reptiles

10. An animal that hunts another animal for food is called a (n) _____.

    prey   vertebrate   predator   herbivore

Lesson #71

1. What layer of soil is pure rock?

    subsoil   topsoil   bedrock

2. The symbol for which type of front is shown?

    _____

3. Put a ✓ next to each statement that is true about ecosystems.

    _____ The ground in the tundra is covered with permafrost.

    _____ The word taiga means *treeless plain.*

    _____ The rainforest gets about 33 feet of rain per year.

4. The plant kingdom is divided into two groups. What are they?

    A) mushrooms and molds
    B) vascular and nonvascular plants
    C) flowering plants and cone-bearing plants
    D) ferns and cone-bearing plants

5. Match these parts of the water cycle with the descriptions.

    A) water vapor forming droplets   _____ evaporation

    B) water turning to vapor   _____ precipitation

    C) water falling to Earth's surface   _____ condensation

6. In a (physical / chemical) change the composition or make up of the substance does change.

Not every plant has flowers. Some have cones instead of flowers. If you remember, one of the groups of vascular plants is cone-bearing plants. **Conifers** are plants that have needle-like leaves and make seeds inside cones. An example of a conifer is a **pine tree**. An adult conifer grows both small and large cones on its branches. When

it's time, the scales of the cone open and the seeds fall to the ground. They eventually sprout in the soil to become seedlings or are eaten by animals.

7. Plants that have needle-like leaves and makes seeds inside of

   cones are called _____.

8. Cone-bearing plants, or conifers, are examples of (vascular / nonvascular) plants.

Use the food chain to answer the questions below.

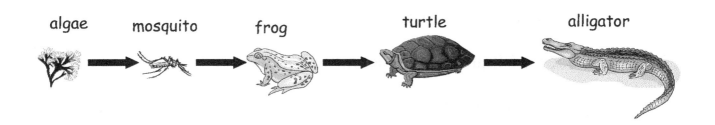

9. Which is the producer? _____

10. A) Which is the final consumer? _____

    B) Where do algae get their energy? _____

Simple Solutions® Science Level 4

Lesson #72

Crossword Puzzle

Word Bank:

erosion	mimicry	evaporation	precipitation
ecosystem	population	rainforest	tundra
taiga	matter	gas	physical

Across

7. All living and nonliving things in an area
9. A group of the same kind of individuals living in an ecosystem
10. The largest ecosystem in the world
11. Imitating the look of another animal in order to stay alive
12. The ground is covered by permafrost

Down

1. Anything that takes up space
2. The process of liquid water becoming a gas
3. The type of change that doesn't change the make up of a substance
4. Does not have a definite shape or take up a definite amount of space
5. Liquid water that falls from the sky as snow, rain, sleet, or hail
6. The process by which the surface of the Earth gets worn down
8. Home to millions of plants and animals

Simple Solutions© Science Level 4

Use the word bank and the clues on the previous page to complete this puzzle.

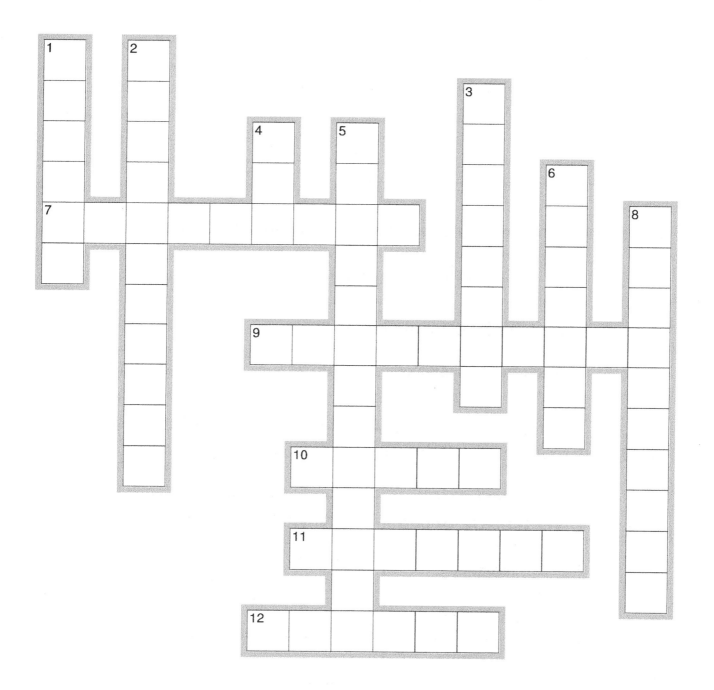

Lesson #73

1. Plants that have needle-like leaves and make seeds inside cones are called _____.

    fungi   conifers   nonvascular   ferns

2. List two items that can be recycled.

    _____

    _____

3. This type of matter takes the shape of its container. What is it?

    solid   liquid   gas

4. Which is an example of an amphibian?

    opossum   snake   turtle   frog   snail

5. Look at the chart below. Write **warm-blooded** or **cold-blooded** next to each vertebrate group. (Hint: Two groups are warm-blooded.)

Vertebrate Group	Warm/Cold-Blooded
mammal	
amphibian	
reptile	
bird	
fish	

6. Which ecosystem is the largest in the world?

    tundra   taiga   grasslands   rainforest   desert

7. Name three examples of liquids.

    _____

    _____

    _____

8. Animals have four basic needs. Name them.

    _____

    _____

    _____

    _____

9. In which stage of metamorphosis is the organism a caterpillar?

    larva   egg   pupa

10. Put a ✓ next to each statement that is true about photosynthesis.

    _____ The food that plants make is sugar.

    _____ The energy for plants to make food comes from the soil.

    _____ Chlorophyll is the green substance inside leaves.

Lesson #74

Use the graph to answer the questions below.

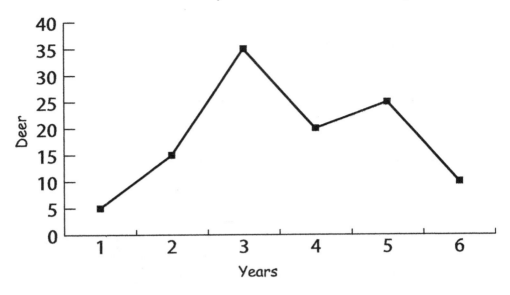

1. How many times was there a decrease in the deer population from one year to the next over the 6 year period?

   _____ times

2. Name something that might have caused this decrease in the deer population.

   _____

3. In what year do you see the greatest increase in the deer population?

   _____

4. Water freezing is an example of a _____.

   A) solid changing to a liquid   C) liquid changing to a solid

   B) gas changing to a solid   D) gas changing to a liquid

Simple Solutions© Science   Level 4

5. Which vertebrate group has hair or fur and produces milk for their young?

       bird   mammal   reptile   amphibian

6. Which form of energy is needed to change water from a liquid to a gas?

       sound   light   heat   chemical

7. Which process caused the rainwater in the bucket to disappear as it sat outside in the sun?

       condensation   precipitation   evaporation   erosion

8. The function of a plant's roots are to take in water and nutrients and _____.

   A) produce fruit   C) anchor the plant
   B) make seeds   D) make food

9. Frogs lay eggs that develop into tadpoles and then develop into adult frogs. This sequence of changes is an example of how living things _____.

   A) make a food chain   C) go through a life cycle
   B) go through photosynthesis

10. What information is contained on this map of the United States?

    A) state capitals
    B) weather conditions
    C) mountain ranges

Lesson #75

1. Kerry uses old newspapers to cover the floor when she paints. This is an example of _____.

    erosion   weathering   recycling   observing

Use this food chain to answer the questions below.

sun → apple tree → field mouse → owl

2. Which organism in this food chain uses the Sun's energy to produce food?

    _____

3. Which organism in this food chain depends directly on plants for food?

    _____

4. Which organism is a carnivore? _____

5. Put a ✓ next to each statement that is true about invertebrates.

    _____ Insects have 6 legs and three body parts.

    _____ Spiders are the largest group of invertebrates.

    _____ Invertebrates have no backbone.

6. Which front symbol is shown?

   _____

7. Which of these is **not** a type of soil?

   loam   sand   larva   silt   clay

8. Choose the correct formula for photosynthesis.

   A) oxygen + water → sugar + carbon dioxide
   B) carbon dioxide + water → sugar + oxygen
   C) carbon dioxide + nutrients + water → sugar

9. Circle the best hypothesis.

   A) Does milk have more vitamins than orange juice?
   B) Milk has more vitamin D than orange juice.
   C) Milk tastes better than orange juice.

10. Decide whether the phrases below describe a **physical change** or a **chemical change**, and then put each under the correct heading in the chart.

    shredding paper   burning a log   baking a cake

Physical Change	Chemical Change

Lesson #76

1. This type of matter has no definite shape and does not take up a definite amount of space. What is it?

    solid   liquid   gas

2. Which of these best describes an animal's instinct?

    A) a butterfly gets nectar from a flower
    B) a dolphin jumps through a hoop
    C) a dog comes when it hears its name

3. Write 1 through 5 to show how energy is passed along in this food chain.

    ____ frog   ____ grass   ____ sun   ____ snake   ____ cricket

4 – 5. Draw the food chain above in the correct order.

6. Match each type of front with its definition.

   _____ stationary front   A) a type of front that stays in one place for many days

   _____ cold front   B) shown on weather maps as a red line with half circles on it

   _____ warm front   C) shown on weather maps as a blue line with triangles on it

7. Which clouds are thin and wispy and look like curls of hair?

   stratus   cirrus   cumulus   cumulonimbus

8. Tyrone wants to look more closely at a cocoon in a jar. Which instrument should he use?

   balance   microscope   hand lens   barometer

9. A tremor or shaking of the earth's surface, usually caused by movement of rock in the crust, is a(n) _____.

   tornado   volcano   landslide   earthquake

10. Write T if the statement is true or F if it is false.

    _____ A) In a physical change, the make-up of the substance changes.

    _____ B) Observing and Asking Questions is the first step of the scientific method.

Lesson #77

Layers of Earth

The Earth contains four layers. If you could cut open Earth, this is what you would find inside. The thin outer layer is called the **crust**. This is the part of the Earth you walk on. The **mantle** is the layer below the crust. Parts of the mantle are so hot that rock melts. This melted rock forms magma. At the center of the Earth is the **core**. The **outer core** is liquid and the **inner core** is solid. The inner core is almost as hot as the sun. It remains solid because there is so much pressure on it.

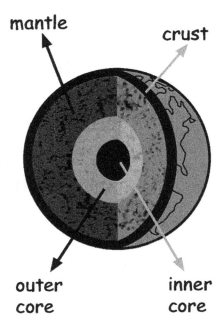

1. List the four layers of Earth.

   _____

   _____

   _____

   _____

2. Plants that have needle-like leaves and make seeds inside cones are called _____.

       fungus   conifers   nonvascular   molds

3. To which group do frogs and salamanders belong?

   A) reptiles   C) insects

   B) vertebrates   D) amphibians

4. Put the steps of the scientific method in the correct order.

    A) Drawing Conclusions

    B) Forming a Hypothesis

    C) Planning an Experiment

    D) Observing and Asking Questions

    E) Conducting an Experiment

    1) _____
    2) _____
    3) _____
    4) _____
    5) _____

5. All of the populations living in the same place make up a(n) _____.

    population   community   ecosystem

6. Which of these items cannot be recycled?

    aluminum cans   newspapers   disposable diapers   cereal boxes

7. Which of these is not a layer of Earth?

    inner core   mantle   clay   crust

8. The first link in any food chain is always the _____.

9. The frozen layer of subsoil in the tundra is called _____.

    erosion   permafrost   bedrock   loam

10. The (taiga / rainforest) is the largest ecosystem in the world.

Lesson #78

1. Which of the following causes erosion?

    A) evaporation

    B) moving water

    C) condensation

2. What is a large body of air called?

    air mass   storm   wind   stationary front

3. The series of changes in appearance that some organisms go through is called _____.

    photosynthesis   metamorphosis   erosion

4. Which two of these do most reptiles, amphibians, and fish have in common?

    A) They have feathers.

    B) They don't have a backbone.

    C) They lay eggs.

    D) They are cold-blooded.

5. Which of these has the greatest mass?

    A) a toothpick

    B) a football

    C) a bottle of laundry soap

    D) a pencil

6. Both plants and animals need air to survive. What part of the air do animals use and what part of the air do plants use?

   plants use _____

   animals use _____

7. What are the four basic needs of plants?

   A) light, carbon dioxide, soil, and water
   B) water, light, shelter, oxygen
   C) light, air, water, nutrients
   D) soil, food, water, carbon dioxide

8. What are the two main groups of plants?

   A) vertebrates and invertebrates
   B) producers and consumers
   C) molds and mushrooms
   D) vascular and nonvascular

9. What do you call an organism that makes its own food?

   an ecosystem   a producer   a consumer

10. What does a balance help you to do?

    A) measure mass
    B) measure temperature
    C) measure wind speed

Lesson #79

Force

Any kind of push or pull is a **force**. A push moves an object away from you and a pull moves an object toward you. When you push a broom you are using a pushing force. You can open a car door by using a pulling force. Any change in motion needs force. You use force to slow down, to speed up, to stop, or to change direction. An object will keep moving until another force stops it. For example, when you catch a football, the force from your hand stops it. One type of force that stops things or slows them down is **friction**. There is more friction between rough surfaces than between smooth or slippery ones.

1. Any kind of push or pull is called _____.

2. A type of force that stops or slows things down is _____.

   gravity   sound   electricity   friction

3. Match these parts of the water cycle with the descriptions.

   A) water turning to vapor   _____ evaporation

   B) water vapor forming droplets   _____ precipitation

   C) water falling to Earth's surface   _____ condensation

4. In the United States, weather patterns generally move from _____ to _____.

5. If a bird flies south each winter, what is it doing?

   evaporating   hibernating   migrating

6. List two traits of fish.

   _____

   _____

7. **Mold is an example of a _____.**

   cone   fungus   fern   cactus

8. Which is a living part of the environment?

   soil   log   oxygen   lobster   rain

9. Which of these would an herbivore eat?

   rabbit   lettuce   worm   robin

10. Read each phrase below. If the phrase describes an insect, write the word **insect** on the line. If it describes a spider, write the word **spider** on the line. If the phrase describes both, write **both** on the line. (See Lesson #17.)

    A) the largest group of invertebrates   _____

    B) makes silk   _____

    C) has 8 legs and 2 body parts   _____

    D) has an outer covering   _____

    E) has 6 legs and 3 body parts   _____

Lesson #80

Gravity

Gravity is a force that pulls objects toward each other. When you throw something into the air, you know it will come back down. The object comes back to you because Earth's gravity pulls on it. Earth's gravity pulls objects toward the center of the Earth. Gravity exists between all objects, not just between objects and Earth. Gravity acts on objects without even touching them.

1. A force that pulls objects toward each other is called _____.

2. Any kind of push or pull is called _____.

   matter   force   gravity   friction

3. Match the layers of the Earth with their definition.

   _____ inner core   A) the layer below the crust

   _____ mantle   B) this layer is solid; it is almost as hot as the sun

   _____ crust   C) under the mantle; this layer is liquid

   _____ outer core   D) the thin outer layer; this is made up of the continents and oceans

4. In a (physical / chemical) change, the make-up of a substance changes.

5. When two different large air masses meet, a _____ is formed.

    front   fault   landslide   stratus cloud

6. A type of force that stops things or slows them down is called _____.

    gravity   friction   wind   erosion

7. A cluster of sunflowers is an example of a(n) _____.

    community   population   ecosystem

8. Which three would most likely be found in the tundra?

    army ants   polar bear   buffalo   Arctic fox   caribou

9. Write T if the statement is true or F if it is false.

    _____ Nonvascular plants do not have tubes.

Use the water cycle diagram to answer the questions below.

10. A) What process is occurring at **B**?

    A) precipitation
    B) evaporation
    C) condensation

    B) What process is occurring at **C**?

    A) precipitation
    B) evaporation
    C) condensation

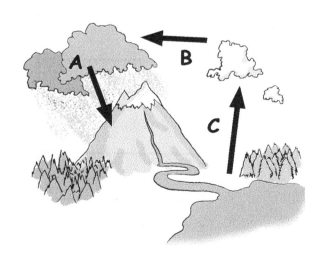

Lesson #81

1 – 2. Complete each sentence with the correct word.

friction   gravity   force

A) A leaf falls off a tree because of the force of _____.

B) _____ will cause an object to slow down, to speed up, to stop, or to change direction.

C) A kickball slows down and eventually stops rolling because of the force of _____.

3. What does this symbol stand for?

   _____

4. Put each word in the correct column of the graphic organizer.

oxygen   dish soap   lamp

milk   carbon dioxide   ice-cube

Solid	Liquid	Gas

5. The process in which a liquid changes to a gas is called _____.

      condensation   precipitation   evaporation

6. _____ is the decayed remains of plants and animals.

      Topsoil   Humus   Subsoil   Permafrost

7 – 8. Label the diagram using the words below. Write each part of the formula of photosynthesis on the line.

      sunlight   carbon dioxide   water   oxygen

A) _____

B) _____

C) _____

D) _____

9. The vent and the cone-like mountain left by the overflow of erupted lava is called a(n) _____.

      volcano   earthquake   tornado   landslide

10. Underline the invertebrates.

      jellyfish   dragonfly   whale   tree frog   octopus   ant

Lesson #82

1. The chart shows two categories. Put each organism in the list in the correct category.

   sunflower   robin   seaweed   hawk   dandelion   blue gill

Producer	Consumer

2. Matter is anything that _____.

   A) stops another object
   B) pulls objects toward each other
   C) pushes or pulls
   D) takes up space

3. Seeds can be spread by wind, by water, and by _____.

   erosion   animals   photosynthesis

4. Which type of matter takes the shape of its container?

   solid   liquid   gas

5. (Chlorophyll / Sugar) is the green substance inside leaves.

Simple Solutions© Science   Level 4

6. The movement of water through the environment is called _____.

   A) precipitation
   B) the water cycle
   C) evaporation

7. A force is a _____ or a _____.

8. Which has slightly more mass, the owl or the apples?

   _____

9. What is the correct order for complete metamorphosis?

   A) pupa → adult → egg → larva
   B) egg → larva → pupa → adult
   C) egg → pupa → larva → adult
   D) larva → egg → pupa → adult

10. Match each word with its definition.

    A) eats only plants   _____ omnivore

    B) eats plants and animals   _____ carnivore

    C) eats other animals   _____ herbivore

165

Lesson #83

1. What three things do plants need for photosynthesis to occur?

   A) carbon dioxide, sunlight, oxygen

   B) sunlight, water, carbon dioxide

   C) sugar, water, carbon dioxide

2. Which of these is the second step of the scientific method?

   A) Planning an Experiment

   B) Forming a Hypothesis

   C) Observing and Asking Questions

3. Give an example of a chemical change.
   (See Lesson #69 if you need help.)

   _____

4 – 5. Read the animal names below. Write each name under the correct group.

| shark | parrot | salamander | snake | turtle |
| human | catfish | frog | whale | penguin |

Amphibian	Fish	Bird

Mammal	Reptile

6. Which part of the plant carries water and nutrients to the leaves?

    roots   stem   flower

7. Match each animal group to its definition.

    _____ Bird   A) has hair or fur; feeds young with milk

    _____ Reptile   B) two legs, wings, and feathers

    _____ Fish   C) moist-skin, lives near water

    _____ Mammal   D) dry, scaly animal that lays eggs

    _____ Amphibian   E) lives whole life in water, breathes with gills

8. An organism with a backbone is called a(n) _____.

    invertebrate   fungus   vertebrate

9. What do you call an organism that eats other living things in order to get energy?

    a producer   a consumer   a food chain

10. What are two traits of birds?

    _____

    _____

Lesson #84

1. Which of these is not a type of precipitation?

    snow   sleet   hurricane   rain   hail

2. Draw the symbol for a cold front in the box.

3. Any kind of push or pull is called _____.

    friction   force   gravity   weathering

4. Which clouds are puffy clouds and sometimes look like pieces of floating cotton?

    stratus   cirrus   cumulonimbus   cumulus

5. Besides clay and silt, list the two other types of soil. (See Lesson #41.)

    _____

    _____

6. Which ecosystem is a cold and treeless plain?

    A) desert   C) rainforest

    B) grassland   D) tundra

7. A type of force that stops or slows things down is _____.

   gravity   sound   electricity   friction

8. Label the parts of a plant.

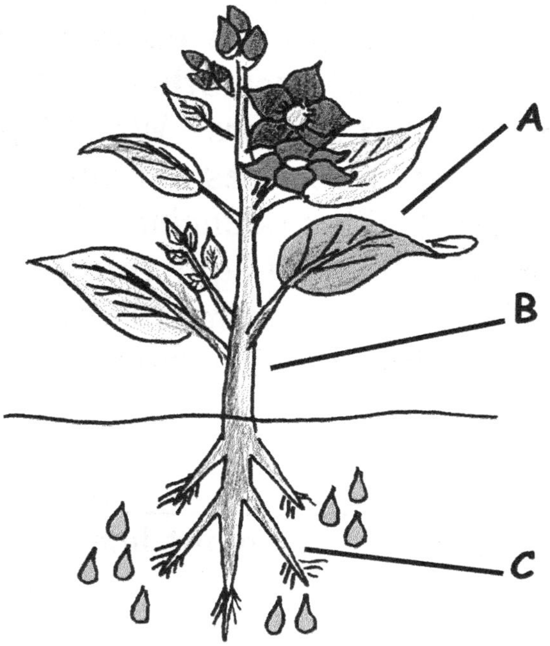

   A) _____

   B) _____

   C) _____

9. In which step of the scientific method do you use your senses to gather information?

   A) Forming a Hypothesis
   B) Observing and Asking Questions
   C) Planning an Experiment

10. For each organism in the chart below, decide whether it is a producer or a consumer, and put a ✓ in the correct column.

	Producer	Consumer
algae		
sea gull		
daisy		

Lesson #85

Light Energy

Light is a form of energy that travels through space. Light travels very rapidly. Nothing travels faster than light, not even sound. Light travels in straight lines until it hits something. It does not travel around corners. When light hits any object, it bounces off the surface of the object. This bouncing off is called **reflection**. Not all light that hits an object is reflected. Sometimes light passes through objects. Our main source of light on Earth comes from the Sun.

1. _____ is a form of energy that travels through space in a straight line.

2. When light hits an object and bounces off the surface; this bouncing off is called _____.

3. What is our main source of light? _____

4. Which two would most likely be found in the desert?

    roadrunner   polar bear   lemming   caribou   tarantula

5. A large body of ice that moves slowly down a slope is a _____.

    glacier   landslide   volcano   sleet

6. (Mimicry / Instinct) is imitating the look of another animal.

Use the chart below to answer the next few questions.

Insect	Diet	Defense
ladybug	aphids (small insects with soft bodies which are found in most yards and in trees)	oozes blood, which has an odor, from its leg joints Some adult lady bugs will fall to the ground and "play dead."
stinkbug	leaves, flowers, fruit, any smaller insect, and caterpillars Many eat sap from plants.	releases an unpleasant odor from 2 glands on its thorax
millipede	decaying vegetation	releases a chemical substance from stink glands to repel predators

7. Which insect eats caterpillars? _____

8. What do all three of these insects have in common?

   _____

9. What do ladybugs eat? _____

10. What is shown here?

    _____

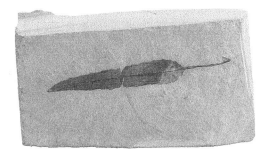

Lesson #86

Transparent/Translucent/Opaque

Not all light that hits an object is reflected. Sometimes light passes through objects. Materials that are **transparent** let most of the light that hits them pass through. Materials that are transparent include glass, water, and some clear plastics. Materials that are **translucent** absorb or scatter some of the light and let the rest pass through. Wax paper, some plastics, and frosted glass (pictured) are translucent materials. Materials that do not allow any light to pass through are called **opaque** materials. An opaque object reflects or absorbs all of the light that reaches it. Some examples of opaque materials include wood, metal, brick, and cardboard.

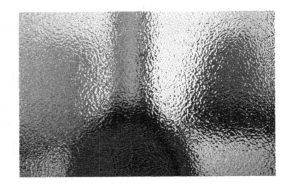

1. Match each word with its definition.

   A) lets most light pass through   _____ opaque

   B) lets some light pass through   _____ transparent

   C) lets no light pass through   _____ translucent

2. When light hits an object and bounces off the surface, this bouncing off is called _____.

   friction   mimicry   reflection   transparent

3. Our main source of light is _____.

   the moon   a lamp   the sun   electricity

Simple Solutions© Science Level 4

4 – 5. Put each word in the correct column of the graphic organizer.

paint   rain   bar soap

helium   ice-cube   oxygen

Solid	Liquid	Gas

6. What layer of soil is best for growing plants?

        subsoil   topsoil   bedrock

7. Which stage of metamorphosis is shown?

    _____

8. Which is **not** a basic need of animals?

        water   wings   food   shelter

9. An animal that is hunted by another animal for food is called _____.

        prey   predator   herbivore   producer

10. Animals without a backbone are called _____.

Lesson #87

1. Put a ✓ next to each statement that is true about light.

   _____ Opaque materials do not allow light to pass through.

   _____ Water and glass are examples of transparent materials.

   _____ Light can travel around corners.

2. Which of these best describes an animal's instinct?

   A) a lion attacks an antelope
   B) a bird builds a nest
   C) a cat uses a litter box

3. Wood, metal, brick, and cardboard are examples of what type of materials?

   translucent   opaque   transparent

4. A force that pulls objects toward each other is called _____.

   weathering   friction   gravity   reflection

5. Which are invertebrates?

   squirrel   ape   clam   ant

   lobster   crab   jellyfish   elephant

6. Which word means *an educated guess*?

   hypothesis   observation   investigation

Simple Solutions© Science   Level 4

7. At the Earth's center is the _____.

   mantle   crust   core

8. Write T if the statement is true or F if it is false.

   _____ A) Matter is anything that takes up space.

   _____ B) Glass bottles cannot be recycled.

9 – 10. Decide whether the phrases below describe a **physical change** or a **chemical change**, and then list each under the correct heading in the chart.

Remember: In a physical change the composition of a substance does not change, and in a chemical change the composition of a substance does change.

burning wood   cutting hair   a chair rusting

melting ice   baking a cake   boiling water

	Physical Change	Chemical Change
A)		
B)		
C)		
D)		
E)		
F)		

Lesson #88

How Do You See Light?

The simplest way to answer this question is this: Light bounces off objects. When the light bouncing off an object reaches your eye, you see it. Let's break this down into steps.

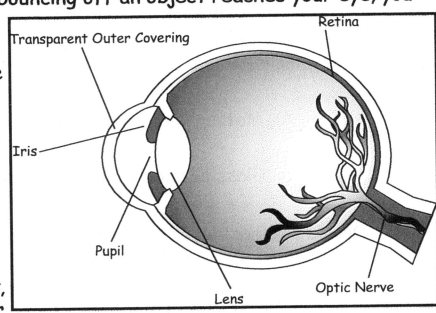

1) The light enters the front of your eye and passes through the **pupil**.
2) The **iris** surrounds the pupil and controls how much light can pass into the eye. When the light is really bright, your iris makes your pupil small. When the light is not that bright, the iris makes the pupil large, which allows more light to pass through.
3) The **lens** of your eye bends the light rays to make them focus on the **retina**, which is at the back of your eye.
4) Your **optic nerve** sends information about the light to your brain.
5) Lastly, your brain tells you what you are seeing. Just think that all of this happens in a split second every time you see something!

1. Which part of your eye controls how much light can pass into the eye?

    optic nerve   pupil   retina   iris

2. Which part of your eye sends the information about the light to your brain?

    optic nerve   pupil   retina   iris

3. Which is the part of the eye where light enters?

    optic nerve   pupil   retina   iris

Simple Solutions© Science Level 4

4. A barometer measures (air pressure / wind speed).

5. Write T if the statement is true or F if it is false.

   _____ Nothing travels faster than light.

6. Fill in the missing steps of the scientific method.

   _____

   Forming a Hypothesis

   _____

   Conducting an Experiment

   _____

7. Draw the recycle symbol in the box.

8. Underline two examples of fungi.

   moss   mushroom   fern   poison ivy   mold

9. Which instrument is shown to the right?

   _____

10. What do you call a dry, scaly animal that lays eggs and lives on land?

    amphibian   mammal   reptile   bird

Lesson #89

1 – 2. Read each type of change below. Write each type of change under the correct heading below.

    ice cube melting   match burning   folding paper

    painting a wall   iron rusting   crushing a can

Physical change	Chemical change

3. During photosynthesis, a gas is given off. Which gas is it?

    carbon dioxide   helium   oxygen

4. What is the role of a producer in a food chain?

    A) eat other animals   C) eat decayed animals

    B) make food   D) eat plants

5. What can happen when an area doesn't get enough rain?

    A) a flood

    B) a drought

    C) a landslide

6. A group of organisms of the same kind make up a _____.

   ecosystem   habitat   population

7. Match each stage of butterfly metamorphosis with its definition.

   _____ larva   A) the caterpillar is wrapped in a cocoon

   _____ egg   B) the life cycle begins here

   _____ adult   C) the egg hatches into a caterpillar

   _____ pupa   D) a butterfly comes out of the cocoon

8. What do you call anything that disguises an animal or helps it hide?

   instinct   hibernate   camouflage

9. What do you call the place where an organism lives in an ecosystem?

   its prey   its instinct   its habitat

10. What makes all animals consumers?

    A) Animals need shelter to survive.
    B) Animals need more energy than plants.
    C) All animals eat plants or other animals.
    D) Animals make their own food.

Simple Solutions© Science   Level 4

Lesson #90

Use the food chain to answer the questions below.

1. Which is an herbivore?   _____

2. Which is the final consumer? _____

Use the Venn diagram to answer questions 3 – 5.

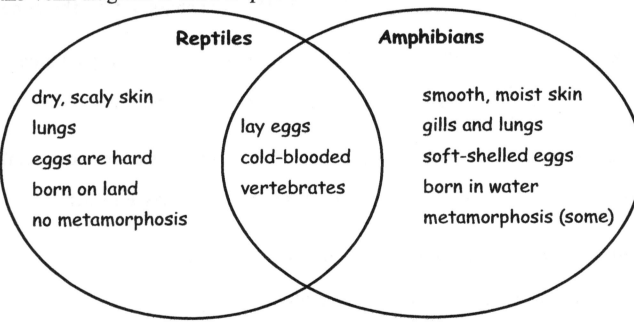

3. What three things do reptiles and amphibians have in common?

   _____

   _____

   _____

4. Using the Venn diagram from the previous page, name the vertebrate group that is born in water.

   _____

5. Using the Venn diagram from the previous page, name the vertebrate group that does **not** go through metamorphosis.

   _____

6. What is the green substance inside leaves called?

   chlorophyll   photosynthesis   sugar   vascular

7. Materials that do not allow any light to pass through are called _____ materials.

   translucent   opaque   transparent

8. The process of water vapor changing to liquid water is _____.

   condensation   precipitation   evaporation

9. _____ is a type of force that stops things or slows things down.

   Friction   Matter   Motion

10. Which ecosystem has the most abundant plant and animal life?

    tundra   rainforest   desert   freshwater

Lesson #91

Sound Energy

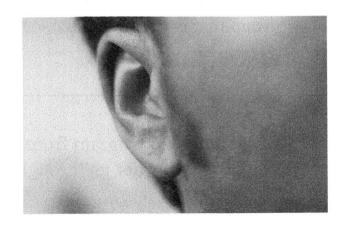

Sound is a form of energy that travels in waves. Sound comes from objects that **vibrate**, or move back and forth. You hear sounds when vibrations move through the air to your ears. When sound waves reach your ears, your eardrum vibrates, and that allows you to hear. When you pluck a guitar, the strings create sound waves in the air around it, and you hear sound. The word **pitch** means *the highness or lowness of a sound*. The short, thin strings on a harp make sounds with a high pitch, and the long, thick strings make sounds with a low pitch.

1. The highness or lowness of a sound is called _____.

2. Another word for *move back and forth* is _____.

3. Materials that allow some light to pass through are called _____ materials.

    translucent   opaque   transparent

4. List the four layers of Earth from outside to inside.

    _____   _____

    _____   _____

5. Which of these layers is the thinnest? _____

6. Plants that have needle-like leaves and make seeds inside of cones are called _____.

   A) fungi
   B) conifers
   C) nonvascular
   D) mosses

7. Which is a **nonliving** part of an environment?

   grasshopper   sunflower   water vapor   wasp

8. Which type of front symbol is shown here?

   _____

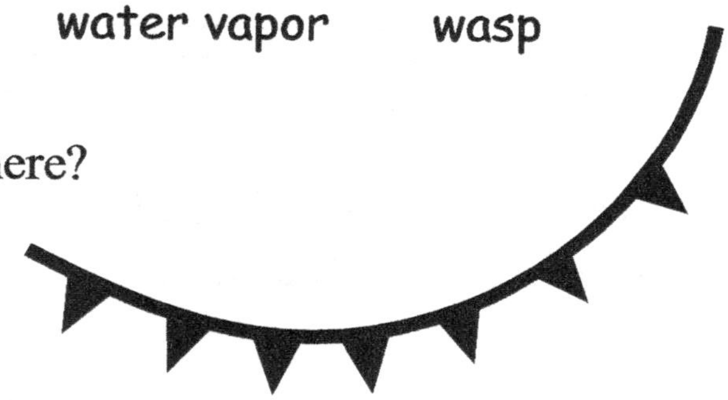

9. Match these parts of the water cycle with the descriptions.

   A) water turning to vapor   _____ condensation

   B) water vapor forming droplets   _____ precipitation

   C) water falling to Earth's surface _____ evaporation

10. In which ecosystem is the ground covered by a layer of frozen subsoil called permafrost?

    A) taiga
    B) rainforest
    C) tundra
    D) grasslands

Lesson #92

Magnets

A **magnet** is a piece of metal that attracts iron and steel. **Attract** means to *pull closer*. A magnet can pull something without even touching it. All magnets have two ends, called **poles** – a **north pole** (N) and a **south pole** (S).

| N | S |

If you put two like poles together (N and N), they **repel** each other. Repel means to *push away*.

If you put two opposite poles together, (N and S) they pull toward each other or **attract**.

1. What do you call the ends of a magnet? _____

2. Another word for *push away* is _____.

3. What material must an object be made from to be attracted to a magnet?

    plastic   iron   rubber   wood

4. Which word means *pull toward*?

    repel   attract   charge

5. Glass, water, and some clear plastics are examples of _____ materials.

    translucent   opaque   transparent

6. In a (physical / chemical) change the composition, or make-up, of the substance does not change.

7. Which type of cloud is shown?

    A) cirrus   C) cumulonimbus

    B) cumulus   D) stratus

8. Which tool measures how hot or cold something is?

    barometer   anemometer   thermometer   balance

9. Write T if the statement is true or F if it is false.

    _____ When light hits an object and bounces off the surface, this bouncing off is called reflection.

10. Which sequence shows forms of water from the hottest to the coldest temperatures?

    A) ice, water vapor, liquid water
    B) water vapor, liquid water, ice
    C) liquid water, water vapor, ice
    D) water vapor, ice, liquid water

Lesson #93

Static Electricity/Current Electricity

Static electricity is an electric charge that builds up on an object. You may see a person's hair stand up on a cold day. You may get a shock when you rub your feet across the carpet and then touch an object or another person. That means that electrical charges have built up on one object and are "jumping" from that object to another. **Lightning** is an example of static electricity.

static

Current electricity is electricity that **moves through wires**. When you plug a toaster into a wall outlet, electricity flows from the wires in the wall through the plug and into the toaster's wires. You use current electricity to heat your home, to light your home, and to cook your food.

1. What do you call an electric charge that builds up on an object?

   _____

2. Electricity that moves through wires is _____.

3. Give one example of static electricity and one example of current electricity.

   static electricity _____

   current electricity _____

4. (Cold fronts / Stationary fronts) stay in one place for many days.

5. Which are two traits of reptiles?

   A) moist skin and scales
   B) dry skin and scales
   C) hair and cold-blooded
   D) dry skin and warm-blooded

6. Label the layers of the Earth.

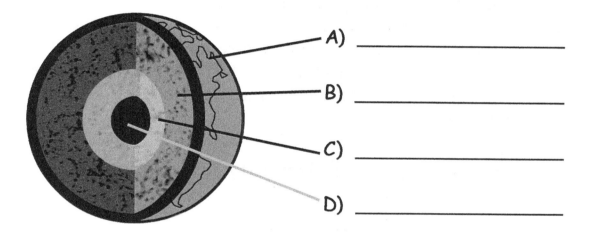

   A) _____
   B) _____
   C) _____
   D) _____

7. The part of the plant that makes food is called _____.

   roots   stem   flower   leaves

8. Choose an organism that goes through metamorphosis.

   squirrel   bird   frog   turtle

9. Which of these is an example of a fungus?

   elm tree   fern   tulip   mushroom

10. This type of matter has a definite shape and takes up a definite amount of space. What is it?

    solid   liquid   gas

Lesson #94

Conductor/Insulator

A conductor is something that lets electricity pass through it easily. Copper and all metals are excellent conductors of electricity. Other examples are aluminum cans, wire, coins, paper clips, and aluminum foil.

An insulator is something that doesn't let electricity flow through it easily. Examples of insulators are plastic, glass, wood, paper, and rubber.

Pure water does not conduct electricity. However, most of the water that we see every day is not pure. Water in oceans, lakes, and swimming pools has minerals dissolved in it; water that comes out of the faucet also has materials dissolved in it. This "impure" water is an excellent conductor of electricity. (In this book, when we say water, we mean the water that you encounter every day; if we mean pure water, we will say so.)

1. Something that lets electricity pass through easily is called a(n) _____.

2. List two examples of a conductor.

   _____

   _____

3. Something that doesn't let electricity pass through easily is called a(n) _____.

4. The green substance inside of leaves is called (sugar / chlorophyll).

Simple Solutions© Science   Level 4

Use the chart to answer the questions below.

Vertebrates	Body Covering	Reproduction	Breathing	Cold/Warm-Blooded
Mammal	hair or fur	live birth	lungs	warm-blooded
Fish	scales	lay eggs in water	gills	cold-blooded
Bird	feathers	lay eggs	lungs	warm-blooded
Amphibian	moist skin	lay eggs in water	gills, then lungs	cold-blooded
Reptile	dry, scaly skin	lay eggs	lungs	cold-blooded

5. Which vertebrate group breathes with gills and then lungs?

   _____

6. Which vertebrate groups are warm-blooded?

   _____

7. Which vertebrate group does not usually lay eggs?

   _____

8. Which two vertebrate groups lay eggs in the water?

   _____ _____

9 – 10. Name two ways reptiles and birds are alike.

   _____ _____

Lesson #95

Current Electricity/Simple Circuit

The path that electricity follows is called a **circuit**. A **simple circuit** consists of a battery, a bulb, and a wire. Often a switch is added to turn the light off and on. A successful circuit is completed when the battery, light bulb, and wire complete a loop, as shown in the picture to the right.

In this picture, the battery is the **source** of the electric current. The positive and negative ends of the battery must be connected to the light bulb by being attached to the bulb's base. Electricity only flows through a completed circuit or a closed loop. The electricity flows through the wire to the light bulb.

1. The path that electricity follows is called a _____.

2. A simple circuit consists of what three main things?

   _____   _____

   _____

3. A force is a _____ or a _____.

4. Which of these has the greatest mass?

   a pencil   a sheet of paper   a computer   a cell phone

5. What does this symbol mean?

   _____

6. Something that doesn't let electricity pass through easily is called a(n) _____.

   conductor   circuit   insulator   static

7. Which clouds are puffy and often look like pieces of floating cotton?

   stratus   cirrus   cumulus   cumulonimbus

8. Circle the best hypothesis.

   A) Fertilizing plants takes too much time.
   B) A plant that is fertilized will grow larger than one that isn't.
   C) All plants need fertilizer.

9. Put a ✓ next to each statement that is true about magnets.

   _____ A) If you put two like poles together, they attract.

   _____ B) The ends of a magnet are called circuits.

   _____ C) An object must be made of iron or steel to be attracted to a magnet.

10. What are two traits of fish?

    _____

    _____

Lesson #96

Work

To a scientist, **work** means *something that is done when a force (push or pull) is used to move an object.* You may feel that reading a novel is work, but to a scientist you haven't done any work at all. You have not physically moved any object. On the other hand, you may feel like you are having fun when you play basketball or kick a football, but a scientist would say you have done work. You have moved the ball.

1. Something that is done when a force moves an object is called _____.

2. Something that lets electricity pass through easily is a(n) _____.

    conductor   circuit   insulator   static

3 – 4. Use these parts to draw a simple circuit in the box below.

 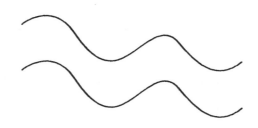 

5. Which is **not** a characteristic of a mammal?

   A) It gives birth to live young.
   B) It has hair or fur.
   C) It is cold-blooded.
   D) It can make milk for its young.

6. The living and nonliving things in an area make up a(n) _____.

       population   ecosystem   habitat

7. What layer of soil lies just below topsoil?

       subsoil   bedrock   humus

8. After the sun, all food chains begin with a _____.

       consumer   producer   omnivore

9 – 10. Look at the words below. Put each word under the correct heading. (See Lesson #94 if you need help.)

   paper   copper   wood   aluminum foil
   coin   plastic   rubber   paper clip

Conductor	Insulator

Lesson #97

Simple Machines/Lever

A **simple machine** is a tool with few or no moving parts that makes work easier. There are **six** simple machines. You will be introduced to each of them over the next six lessons.

If you have to clean the garage, one way to get rid of the dirt is to use a broom. A broom is a simple machine with no moving parts. To use a broom, you push or pull on it with one hand and apply only one force. A broom is a **lever**. A lever is a bar that pivots, or turns, on a fixed point. The fixed point is called the **fulcrum**. When you use a broom, one hand holds the end of the handle. That hand stays still. It is the fulcrum. The other hand pulls the middle of the handle. The broom sweeps up the dirt. Other examples of levers are a **shovel**, a **rake**, a **seesaw**, a **wheelbarrow** and a **hockey stick**.

1. Name two examples of a lever.

   _____   _____

2. Put a ✓ next to each statement that is true.

   _____ A seesaw is a lever.

   _____ The fixed point on a lever is the fulcrum.

   _____ A simple machine has few moving parts.

3. Write 1 through 4 to show how energy is passed along in this food chain.

   ____ frog   ____ caterpillar   ____ dandelion   ____ snake

4. In the food chain in item 3, which is the producer?

   _____

5. An ice-cube melting is an example of a _____.

   A) solid changing to a liquid
   B) gas changing to a solid
   C) liquid changing to a solid
   D) gas changing to a liquid

6. Write T if the statement is true or F if it is false.

   _____ The energy that plants need for photosynthesis comes from the sun.

7. List the three states of matter.

   _____   _____   _____

8. A mushroom is an example of a _____.

   flower   fungus   fern   cactus

9. Which ecosystem gets up to 33 feet of rain per year?

   taiga   rainforest   grasslands   tundra   deciduous forest

10. Which of these would a carnivore eat?

    seeds   dandelions   worms   grass

Lesson #98

Simple Machine/Pulley

A **pulley** is a wheel with a groove that allows a rope or chain to move around it. You pull one end of the rope or chain one way, and the other end moves in the opposite direction. A pulley makes your work seem easier because it works with gravity to change the direction of motion. For example: Let's say you wanted to lift something heavy to the second floor of your barn. You could attach a pulley somewhere on the second floor, take a long piece of rope, and tie one end of the rope around the object you want to lift; loop the rest of the rope through the pulley and let the rest fall down to the floor. Stand on the ground floor and pull straight down on the rope. As you pull down, the object moves into the air until it reaches the second floor. The work seems easier because you are using gravity to help you out. Pulleys are also found on **window blinds, sailboats, cranes,** and **flagpoles**.

1. A wheel with a groove that allows a rope to move around it is called a _____.

2 – 3. When using a pulley, why does the work seem easier?

_____

_____

4. A _____ is a tool with few or no moving parts that makes work easier.

5. The path that electricity follows is called a _____.

    circuit   current   fuse   conductor

6. Look at the chart below. Read the description of each ecosystem. Write the name of the ecosystem next to its description.

    tundra   rainforest   taiga

Description	Ecosystem
abundant plant and animal life; found near the equator	
largest ecosystem in the world	
cold and treeless; permafrost found here	

7. Electricity that moves through wires is _____ electricity.

    static   heat   current   motion

8. A plant produces _____ through photosynthesis and then uses it for energy.

    water   oxygen   sugar   carbon dioxide

9. In which stage of metamorphosis is the organism wrapped in a cocoon?

    larva   egg   pupa

10. A broom is an example of a (lever / pulley).

Lesson #99

Simple Machine/Inclined Plane

An **inclined plane** is a simple machine that makes lifting and moving things a lot easier. If you are trying to put heavy furniture onto the back of a truck and it's too heavy to lift, what can you do? You can get a long board and put one end on the ground; then lean the other end onto the back of the truck to make a **ramp**. A ramp is an inclined plane. Now you can push the heavy furniture up the ramp and onto the truck. The inclined plane made your job a lot easier.

1. A simple machine that makes lifting and moving things easier is called a(n) _____.

2. An example of an inclined plane is a _____.

3. Match each of the simple machines you've learned so far with its description.

   _____ lever   A) it makes lifting and moving things easier

   _____ pulley   B) a bar that pivots on a fixed point

   _____ inclined plane C) a wheel with a rope around it

4. Give two examples of a gas.

   _____   _____

5. Which step of the scientific method comes after **Forming a Hypothesis**?

   A) Drawing Conclusions
   B) Planning an Experiment
   C) Conducting an Experiment

6. Dew forming on a leaf is the result of which process?

   condensation   evaporation   erosion

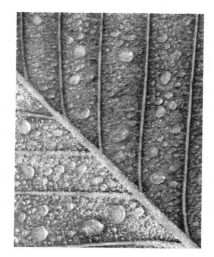

7. What three things are needed for photosynthesis to occur?

   A) sunlight, sugar, and oxygen
   B) carbon dioxide, sugar, and sunlight
   C) carbon dioxide, water, and sunlight

8. Most earthquakes occur along _____.

   faults   hills   oceans   rocks

9. Which of these is a spinning column of air that touches the ground?

   blizzard   tornado   drought

10. What makes up humus?

    A) fertilizer
    B) minerals
    C) decayed plants and animals
    D) rocks

Lesson #100

Simple Machine/Wheel-and-Axle

A **wheel-and-axle** is made up of a small cylinder or an axle attached to a larger wheel. The wheel-and-axle are connected so that they turn together. One example of a wheel-and-axle is a **doorknob**. The knob part of the door is a wheel. The rod connected to it, which goes through the door to the other side, is an axle. When you turn the knob, the axle turns and the door opens. The knob and the rod are connected and turn together. Other examples of a wheel-and-axle are a **screwdriver**, a **faucet handle**, and a **steering wheel**. Actually, anything with wheels falls into this category.

1. Anything with wheels is an example of which type of simple machine?

   _____

2. The fixed point on a lever is called the _____.

       pulley   fulcrum   friction

3. Vascular plants have _____.

       disease   no roots   tubes   fungus

4. A force that pulls objects toward each other is called _____.

       erosion   gravity   friction   sound

5. The series of changes from birth to adulthood is called _____.

       photosynthesis   metamorphosis   simple circuit

6. List the five groups of vertebrates.

   _____

   _____

   _____

   _____

   _____

7. Which instrument is shown?

   barometer   hand lens   anemometer

8. Which are invertebrates?

   cardinal   butterfly   squid   whale   wasp

9. Name two ways seeds can spread.

   _____

   _____

10. Which part holds up the plant?

    leaf   root   stem   seed

Lesson #101

Simple Machine/Screw

A **screw** is a simple machine that you turn to hold two or more objects together or to lift an object. You use screws to attach boards and other objects. A screw is like a nail with threads around it. When you put a nail in a board, it goes straight through the wood. When you use a screw, it moves around and around as it goes into the wood.

1. A simple machine that holds objects together is a _____.

2. Which of these is not like the others?

    paint   water   ice cube   pancake batter

3. The items that are alike in item 2 are all _____.

    The one that's different is a _____.

4. Which layer of the Earth do you walk on?

    mantle   inner core   crust   outer core

5. _____ is anything that takes up space.

    Work   Friction   Matter   Gravity

6. When two different large air masses meet, a _____ is formed.

    front   hurricane   funnel cloud   fault

7. How long does it take the moon to go through its phases?

    about 20 days   about 50 days   about 30 days

8. Match each cloud type with its description.

    _____ cirrus   A) uniform grayish clouds that often cover the entire sky

    _____ cumulus   B) thin, wispy clouds; they look like wisps of hair

    _____ stratus   C) associated with powerful thunderstorms

    _____ cumulonimbus   D) puffy clouds that sometimes look like pieces of floating cotton

9. The first link in any food chain is always the _____.

10. Match each ecosystem with its definition.

    _____ rainforest   A) cold and treeless

    _____ taiga   B) home to millions of plants and animals

    _____ tundra   C) the largest ecosystem in the world

Lesson #102

Simple Machine/Wedge

A **wedge** is a simple machine made up of two inclined planes placed back to back. You use a wedge to force two things apart or to split one thing into two things. A wedge has a pointed end and a wide end. Some examples of wedges are a **doorstop**, an **axe**, a **chisel**, and a **knife**.

1. A simple machine made up of two inclined planes that is used to force things apart is a _____.

2. If you put two opposite poles of a magnet together, what will happen?

   A) They will attract.

   B) They will repel.

   C) Nothing will happen.

3. Look at the words below. Put each word under the correct heading.

   bean plant   rabbit   daisy   grass   kangaroo

Producer	Consumer

4. What provides the energy for photosynthesis to happen?

    carbon dioxide   oxygen   sunlight   water

5. Which one would be found in the tundra?

    ape   Arctic fox   toucan   roadrunner

6. Which kind of electricity lights your home?

    static electricity   current electricity

7. A ramp is an example of a(n) _____.

    pulley   lever   inclined plane   wedge

8. What are the ends of the magnets called?

    charge   circuit   electricity   poles

9. Arrange the steps of the scientific method in the correct order.

    A) Conducting an Experiment   1. _____

    B) Observing and Asking Questions   2. _____

    C) Drawing Conclusions   3. _____

    D) Planning an Experiment   4. _____

    E) Forming a Hypothesis   5. _____

10. A (producer / consumer) makes its own food.

Lesson #103

1. Which term includes herbivores, carnivores, and omnivores?

   A) producers
   B) prey
   C) consumers
   D) vascular

2. Which clouds are grayish and often cover the entire sky?

       cumulus   stratus   cirrus   cumulonimbus

3. What word means *the colors and patterns an animal uses to disguise itself*?

       hibernate   migrate   camouflage

Look at the weather map and use it to answer the questions below.

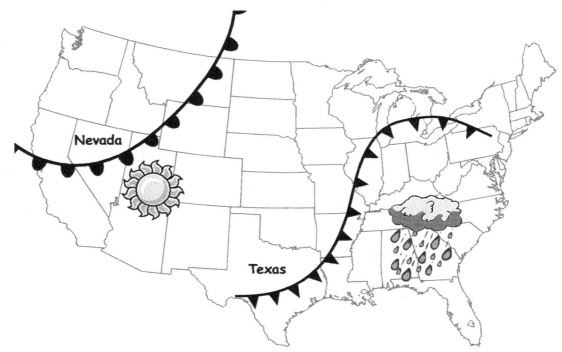

4. Which state is experiencing a cold front?

       Nevada   Texas

5. In which part of the country is it raining?

　　the southeast　　　the northwest　　　no part is having rain

6. Which phase of the moon is shown?

　　A) full moon
　　B) half moon
　　C) crescent moon

7. Which are insulators?

　　glass　　paper　　aluminum　　rubber　　wood

8. Put a ✓ next to each statement that is true about simple machines.

　　_____　A doorstop is a wheel-and-axle.

　　_____　The fixed point on a lever is the fulcrum.

　　_____　A ramp is an example of a pulley.

9. A balance measures _____.

　　size　　temperature　　mass　　speed

10. A large body of ice that moves slowly down a slope is called a ____.

　　glacier　　landslide　　earthquake　　sleet

Lesson #104

1 – 2. Read the description of each simple machine. Match each simple machine to its description. Two choices will not be used.

lever   pulley   inclined plane

wedge   screw   wheel-and-axle

Description	Simple Machine
a rope that is fitted around a fixed wheel; a crane is an example	
made up of two inclined planes; a door-stop is an example	
a bar that pivots on a fixed point; a rake is an example	
makes lifting and moving things a lot easier; a ramp is an example	

3. Where do hurricanes get their energy?

     from the wind   from the sun   from the warm water

4. The green substance inside leaves is called _____.

     chlorophyll   photosynthesis   sugar

5. Which of these is not a type of soil?

     A) clay
     B) silt
     C) moss
     D) sand

6. Something that lets electricity pass through easily is a(n) _____.

      conductor   insulator   static electricity

7. The process where a liquid changes to a gas is called _____.

      precipitation   evaporation   condensation

8. Write T if the statement is true or F if it is false.

    A) _____ Aluminum foil is an insulator.

    B) _____ Work is anything that is done when a force moves an object.

9. All of the populations living in the same place make up a(n) _____.

      population   community   ecosystem

10. Read the description of behaviors that help animals survive. Use the words below to fill in the chart. One word will not be used.

    mimicry   hibernate   migrate   instinct

Description	Behavior
to go into an inactive state during the winter	
to travel from one place to another and back again	
imitating the look of another animal	

Lesson #105

Solar System

The **solar system** consists of a group of objects in space that revolve around the sun. The sun is the center of our solar system. The other objects in our solar system include: planets, moons, asteroids, comets, meteors, rocks, and gases. An **asteroid** is made up of rock and metal. Most asteroids can be found orbiting the sun in a belt that lies between Jupiter and Mars. **Comets** are made of ice, rock, and frozen gases. Comets have very long tails. A **meteor** is also known as a "shooting star." Meteors are usually very tiny. They can be as small as a grain of sand or as large as a softball.

1. Name four objects that make up our solar system.

   _____

   _____

   _____

   _____

2. Most asteroids can be found between which two planets?

   Venus and Mars   Earth and Mars   Jupiter and Mars

3. Which of these is also known as a "shooting star?"

   meteor   comet   asteroid   moon

4. Which phase of the moon is shown?

    full moon   half moon   crescent moon

5. Name the three states of matter.

_____ _____ _____

6. How long does it take the moon to go through its phases?

    about 21 days   about 18 days   about 30 days

7. Plants that have needle-like leaves and make seeds inside of cones are called _____.

   A) nonvascular
   B) mosses
   C) conifers
   D) fungi

8. What form does water take at room temperature?

    solid   liquid   ice   vapor

9. Kevin rubs a balloon on his shirt and he sticks it to the wall. What type of electricity is being used here?

    static electricity   a circuit   current electricity   magnetic

10. Which is a living part of the environment?

    stone   fog   helium   root   snow

Lesson #106

Solar System/Inner Planets

The **inner planets** are the planets closest to the sun. The inner planets include **Mercury, Venus, Earth**, and **Mars**. These planets are smaller than the other planets and are made up mostly of rock. Each one has few or no moons.

Only the planet **Earth** has liquid water on the surface and oxygen in its atmosphere. Because of this, Earth is the only planet that supports life. **Mercury** is about the same size as Earth's moon. It is covered with craters. Mercury is the closest planet to the sun. **Venus** is about the same size as Earth. The surface of Venus has mountains, craters, and volcanoes. **Mars** is called "The Red Planet" because from Earth, it appears to be red. It is smaller than Earth and has huge dust storms. These storms can last for months.

1. Name the four inner planets.

    _____   _____

    _____   _____

2. Which planet is almost the same size as Earth? _____

3. Which planet is called "The Red Planet?" _____

4. (Cirrus / Stratus) clouds are thin and wispy.

Simple Solutions© Science   Level 4

5. Which of these is also known as a "shooting star?"

   meteor   comet   asteroid   moon

6. What is made of ice, rock, and frozen gases and has a long tail?

   meteor   comet   asteroid   moon

7. Something that doesn't let electricity pass through easily is called a(n) _____.

   conductor   insulator   static electricity

8. Match each simple machine with its description.

   _____ lever   A) it makes lifting and moving things easier

   _____ pulley   B) a doorknob is an example

   _____ inclined plane   C) a bar that pivots on a fixed point

   _____ wheel-and-axle   D) a wheel with a rope around it

9. How will these magnets react to each other as they get closer?

   | N   S | | N   S |

   attract   repel

10. What do you call the part of the plant that grows underground and takes water and nutrients from the soil?

    stem   leaf   root   flower

Lesson #107

Solar System/Outer Planets

Saturn

The **outer planets** are the planets farthest from the sun. The outer planets include **Jupiter, Saturn, Uranus,** and **Neptune.** These planets are very large and made up mostly of gases. They are often called the "gas giants." The outer planets have many moons and they are also surrounded by rings which are made of ice, dust, or rock. Pluto was listed as the ninth planet until 2006. At that time, scientists changed their definition of a planet, and since Pluto did not fit the new definition, they now classify Pluto as a "dwarf planet."

Jupiter is the largest planet in the solar system. Jupiter has a giant storm that has been present for more than 300 years. This storm is called the Great Red Spot. **Saturn** is known for its beautiful rings. Saturn has 31 moons. **Uranus** is a planet that rotates on its side. Uranus has about 27 moons. **Neptune** is one of the windiest planets in the solar system. The winds on Neptune can reach up to 1,200 mph.

1. Name the four outer planets.

   _____   _____

   _____   _____

2. Which former planet is now a dwarf planet?

   _____

3. Which outer planet has a large storm called the Great Red Spot?

   _____

4. Which planet is the windiest? _____

5. Which simple machine is made up of two inclined planes? A doorstop is an example.

   pulley   lever   wedge   wheel-and-axle

6. What can happen when an area gets too much rain?

   A) a flood
   B) a drought
   C) a landslide

7. Anything that takes up space is called _____.

   work   friction   matter   electricity

8. What is a hypothesis?

   an experiment   a logical guess   a test

9. What does this picture represent?

   evaporation   precipitation   condensation

10. This type of matter has no definite shape. What is it?

    solid   liquid   gas

Lesson #108

1. Which is the first step of the scientific method?

   A) Observing and Asking Questions

   B) Forming a Hypothesis

2. Match each inner planet with its description.

   _____ Mercury   A) the only planet to support life

   _____ Venus   B) this planet is about the same size as Earth; the surface has mountains, craters, and volcanoes

   _____ Earth   C) called the Red Planet

   _____ Mars   D) the closest planet to the sun

3. In a food chain, consumers _____.

   A) are the first link

   B) use the sun to make food

   C) eat other living things

4. An example of a reptile is a _____.

       moose   toad   lizard   salmon

5. Invisible water in the air is called _____.

       dew   water vapor   rain   ice

6. Wearing away or breaking up old rocks is _____.

    friction   recycling   weathering

7. Which simple machine is a bar that pivots on a fixed point? A broom is an example.

    A) pulley
    B) lever
    C) wedge
    D) wheel-and-axle

8. What is the center of a hurricane called?

    funnel   precipitation   eye   spiral

9. List two examples of an insulator.

    _____

    _____

10. Decide whether each phrase below describes a **physical change** or a **chemical change**. Then put each under the correct heading in the chart.

    baking a cake   roasting marshmallows   painting a wall

Physical Change	Chemical Change

Lesson #109

Look at the graphic organizer. Use it to answer the questions below.

Fruit	Shape	Color	Size
apple	round	green	medium
grape	round	green	small
lemon	oval	yellow	medium
banana	crescent	yellow	medium

1. Which object is small, round, and green? _____

2. How do lemons differ from bananas?

    shape   size   color

3. How do grapes differ from apples?

    shape   size   color

4. Which tool is used to measure wind direction?

    A) weather vane
    B) balance
    C) thermometer

5. Which of these is **not** something most plants need in order to live?

    nutrients   water   soil   air   light

Simple Solutions© Science — Level 4

6. Which layer of soil is best for growing plants?

    topsoil   bedrock   subsoil

7. How does energy move through an ecosystem?

    A) producer ⟶ consumer ⟶ sun
    B) sun ⟶ producer ⟶ consumer
    C) sun ⟶ consumer ⟶ producer

8. Which planet is closest to the sun?

    Venus   Mercury   Mars   Earth

9. Fill in the missing steps of the scientific method.

    1. <u>Observing and Asking Questions</u>

    2. <u>Forming a Hypothesis</u>

    3. _____

    4. _____

    5. <u>Drawing a Conclusion</u>

10. Which of these is an example of a fungus?

    A) moss
    B) fern
    C) mushroom
    D) elm tree

Lesson #110

1. Which clouds are associated with powerful thunderstorms?

    cirrus   stratus   cumulonimbus   cumulus

2. Name the four outer planets.

    J _____

    S _____

    U _____

    N _____

3. Which are conductors?

    paper clip   paper   aluminum   rubber   copper

4. Match each of these objects in the solar system with its description.

    ____ comet   A) found orbiting the sun in a belt that lies between Jupiter and Mars

    ____ meteor   B) made of ice, rock, and frozen gases

    ____ asteroid   C) known as a shooting star

5. Read each statement. If the statement describes a **physical change**, write a P on the line. If the statement describes a **chemical change**, write the letter C.

    ____ baking a cake   ____ cutting your hair

    ____ an icicle melting   ____ lighting a match

6. Which former planet is now called a "dwarf planet?" _____

Use the food chain to answer the questions below.

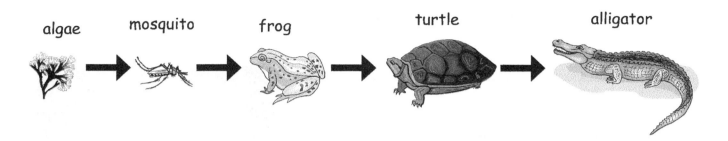

7. In this food chain, which is the producer? _____

8. In this food chain, which animals eat other animals?

   _____

9. After each animal, write the group to which each belongs. One has been done for you.

   A) mosquito   _____

   B) alligator   _____

   C) frog   _____

   D) turtle   _____reptile_____

10. Match each front with its picture.

    _____ warm front

    _____ cold front

221

Lesson #111

1. Something that is done when force moves an object is called _____.

    friction   matter   gravity   work

2. Name the four inner planets.

    M _____

    V _____

    E _____

    M _____

3 – 4. Match each simple machine to its description.

lever   screw   inclined plane
wedge   pulley   wheel-and-axle

Description	Simple Machine
makes lifting and moving things a lot easier; a ramp is an example	
a rope that is fitted around a fixed wheel; a crane is an example	
a bar that pivots on a fixed point; a rake is an example	
made of a small axle and a wheel that turns together	
made up of two inclined planes; a doorstop is an example	
turned to lift an object or to hold two or more objects together	

5. Where do hurricanes get their energy?

   from the sun   from warm water   from the wind

6. The path that electricity follows is called a _____.

   current   fuse   circuit   conductor

7. When light hits an object and bounces off the surface, this bouncing off is called _____.

   A) translucent
   B) reflection
   C) chemical change
   D) opaque

8. What is shown here?

   _____

9. Which is **not** a basic need of animals?

   water   feet   shelter   food

10. Which process is demonstrated in this picture?

    A) condensation
    B) precipitation
    C) evaporation

Lesson #112

1. Which form of energy is needed to change water from a liquid to a gas?

    sound   light   heat   chemical

2. Choose the invertebrates.

    whale   starfish   octopus   zebra

    bat   tarantula   catfish   dragonfly

3. Which has the most mass?

    a twig   a dog   an ant   a popsicle stick

4. What is the source of energy for plants?

    soil   water   sunlight   animals

5. Put a ✓ next to each statement that is true about the water cycle.

    _____ Clouds form when water condenses.

    _____ The water cycle has no beginning or end.

    _____ Water recycles itself.

6. What is the correct order for complete metamorphosis?

    A) egg → pupa → larva → adult
    B) egg → larva → pupa → adult
    C) larva → egg → pupa → adult
    D) pupa → adult → egg → larva

Simple Solutions® Science Level 4

7. Name the eight planets in the order of their distance from the sun. Some beginning letters have been given.

1) __M_____   5) _____

2) _____   6) __S_____

3) __E_____   7) _____

4) _____   8) __N_____

8. Look at the words below. Decide which shows an example of static electricity and which shows an example of current electricity. Write S if it shows **static electricity** and C if it shows **current electricity**.

_____ A) rubbing a balloon on the carpet

_____ B) using a toaster

_____ C) socks sticking together in the dryer

_____ D) using a lamp

9. Write T if the statement is true or F if it is false.

____ A) Wood is an insulator.

____ B) Matter is the force that pulls objects toward each other.

10. Water evaporating is an example of a _____.

A) solid changing to a liquid   C) liquid changing to a solid
B) liquid changing to a gas   D) gas changing to a liquid

Lesson #113

1. To which group do lobsters and crabs belong?

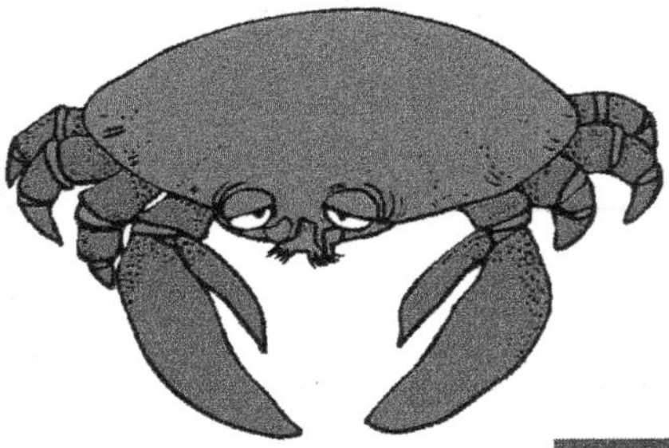

    A) reptiles
    B) invertebrates
    C) insects
    D) mammals

2. Which cloud is shown here?

    A) cumulus   C) stratus
    B) cirrus   D) cumulonimbus

3. An example of a mammal is a _____.

    cardinal   opossum   snake   salamander

4. Plants make their own food through a process called _____.

    A) weathering
    B) metamorphosis
    C) photosynthesis
    D) erosion

5. _____ is the by-product of photosynthesis.

    A) Carbon dioxide
    B) Oxygen
    C) Sugar
    D) Chlorophyll

Simple Solutions® Science   Level 4

6. A ramp is an example of which kind of simple machine?

   lever   pulley   inclined plane   wedge

7. What are the ends of magnets called?

   A) circuits
   B) poles
   C) switches
   D) caps

8. Label the layers of the Earth.

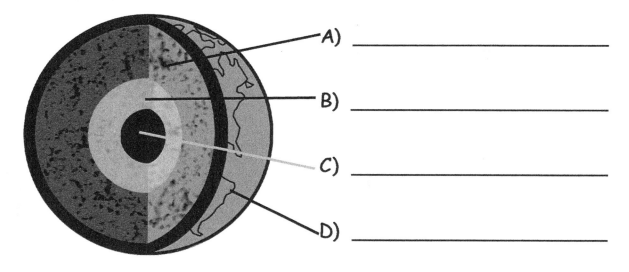

A) _____

B) _____

C) _____

D) _____

9. List two traits of birds.

   _____

   _____

10. A simple machine that holds objects together is called a _____.

    lever   pulley   screw   wedge

Simple Solutions© Science — Level 4

Lesson #114

1. Which phase of the moon is shown here?

   _____

2. A break in the crust where rock moves is called a _____.

   wave   hole   fault   volcano

3. Which circuit will light when the switch is closed? _____

4. An organism without a backbone is called a(n) _____.

   invertebrate   fungus   vertebrate

5. A crane uses which type of simple machine?

   A) lever   C) wheel-and-axle

   B) pulley   D) wedge

6. Name two forms of precipitation.

   _____   _____

7. A polar bear hunts and eats a seal. In this situation, the polar bear is the _____.

           predator   producer   prey

8. If a bird flies south each winter, what is it doing?

   A) migrating
   B) evaporating
   C) hibernating

9. Match each tool with its description.

   _____ hand lens   A) magnify, or make something larger

   _____ thermometer   B) measure the mass of an object

   _____ anemometer   C) measure how hot or cold something is

   _____ balance   D) measure wind speed

10. The ground here is covered by a layer of frozen subsoil called permafrost.

    A) taiga
    B) rainforest
    C) tundra

Lesson #115

1 – 2. Put each animal name under the correct heading in the chart.

spider   raccoon   monkey   crab   worm

salmon   beetle   lizard   jellyfish

Invertebrates	Vertebrates

3. What are the four basic needs of plants?

   A) light, air, water, nutrients
   B) light, carbon dioxide, soil, and water
   C) water, light, shelter, oxygen
   D) soil, food, water, carbon dioxide

4. What layer of soil is pure rock?

   subsoil   topsoil   bedrock

5. What does *vascular* mean?

   with tubes   with cones   with flowers   with soil

6. Choose the vertebrate group(s) that are cold-blooded.

   A) mammals
   B) reptiles
   C) fish
   D) birds
   E) amphibians

7. The living and nonliving things in an area make up a(n) _____.

   ecosystem   organism   water cycle

8. Show how energy is passed along in this food chain from first to last using numbers 1 – 4.

   ____ mouse   ____ grain   ____ hawk   ____ snake

9 – 10. Draw the food chain above in the correct order.

Lesson #116

1. Put each planet under the correct heading in the chart.

   Saturn   Mars   Neptune   Earth

   Uranus   Venus   Jupiter   Mercury

Inner Planets	Outer Planets

2. A doorknob is an example of a(n) _____.

   pulley   inclined plane   wheel-and-axle   lever

3. Lightning is an example of _____.

   current electricity   static electricity

4. Which type of cloud is the highest in the sky and looks like wisps of hair?

   cumulus   cirrus   stratus

5. What word means *to go into a sleep-like state* during the winter months?

   camouflage   hibernate   migrate   instinct

6. Which is **not** a living thing?

    fern   soil   robin   catfish

7. Materials that let some light pass through are called _____.

    opaque   transparent   translucent

8. What is made of ice, rock, and frozen gases and has a long tail?

    meteor   comet   asteroid   moon

9. Water freezing is an example of a _____.

    A) solid changing to a liquid
    B) gas changing to a solid
    C) liquid changing to a solid
    D) gas changing to a liquid

10. For each item in the chart below, decide whether it is a producer or a consumer, and put a ✓ in the correct column.

	Producer	Consumer
grain		
polar bear		
groundhog		
algae		
corn stalk		

Lesson #117

1 – 2. Which circuit will work when the switch is closed? Explain.

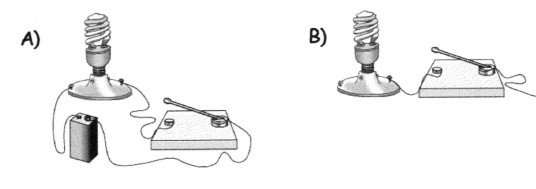

_____

_____

3. Glass and water are examples of which type of materials?

        translucent   opaque   transparent

4. Draw the symbol for recycle.

5. Write whether each item is a **conductor** or an **insulator**.

Item	Conductor/Insulator
copper wire	
cardboard	
penny	
plastic	

6. Match each simple machine with its description.

    ____ lever   A) it makes lifting and moving things easier

    ____ pulley   B) holds things together

    ____ inclined plane C) a bar that pivots on a fixed point

    ____ wheel-and-axle D) a doorknob is an example

    ____ wedge   E) a wheel with a rope around it

    ____ screw   F) two inclined planes placed back to back

7. Any kind of push or pull is called _____.

    gravity   matter   friction   force

8. If you put two like poles of a magnet together, what will happen?

    A) They will attract.
    B) They will repel.
    C) Nothing will happen.

9. A large body of ice that moves slowly down a slope is called a ____.

    snowball   volcano   glacier   landslide

10. What is the part of the plant that grows out of the stem? It is where the plant makes food.

    root   stem   leaf   flower

Lesson #118

1. Name the three states of matter.

   _____

   _____

   _____

2. Which type of cloud is shown to the right?

   A) cirrus   C) cumulonimbus

   B) cumulus   D) stratus

3. Which of these would a carnivore eat?

   grass   lettuce   weeds   trout

4. What is the center of a hurricane called?

   funnel   precipitation   eye   spiral

5. List the five vertebrate groups.

   _____

   _____

   _____

   _____

   _____

6 – 7. Look at the chart below. Read the description of each ecosystem. Write the name of the ecosystem that matches its description.

desert   tundra   taiga   deciduous forest   rainforest

Description	Ecosystem
cold and treeless; permafrost found here	
abundant plant and animal life; found near the equator	
less than 10 inches of rain each year; limited plant life	
largest ecosystem in the world	
trees lose their leaves in the fall	

8. Choose the correct formula for photosynthesis.

   A) oxygen + water → sugar + carbon dioxide

   B) carbon dioxide + nutrients + water → sugar

   C) carbon dioxide + water → sugar + oxygen

9. Materials that allow some light to pass through are called _____ materials.

   translucent   opaque   transparent

10. Animals with a backbone are called _____.

Simple Solutions© Science

Level 4

Lesson #119

1 – 2. Use the word bank to complete the sentences below.

    sugar   water   photosynthesis
    sunlight   carbon dioxide   oxygen

During _____, plants take a

gas called _____ from the air; they pull

_____ up through their roots, and use _____

to make _____. Plants give off _____

as a by-product.

3. Plants that have needle-like leaves and make seeds inside cones are called _____.

         ferns   conifers   nonvascular   fungi

4. Match each type of front with its definition.

   ____ stationary front   A) shown on weather maps as a blue line with triangles on it

   ____ cold front   B) shown on weather maps as a red line with half circles on it

   ____ warm front   C) a type of front that stays in one place for many days

5. When two different large air masses meet, a _____ is formed.

   hurricane	front	fault	cloud

6. What is the role of a producer in a food chain?

   A) eat plants	C) make food
   B) eat decayed animals	D) eat other animals

7. Which kind of change does not create something new?

   physical	chemical

8. Fill in the missing steps of the scientific method.

   1. _____

   2. _Forming a Hypothesis_____

   3. _____

   4. _Conducting an Experiment____

   5. _____

9. Name two forms of precipitation.

   _____   _____

10. How long does it take the moon to go through its phases?

    about 45 days	about 30 days	about 19 days

Lesson #120

1. Write whether each item is **transparent, translucent,** or **opaque**. (See Lesson #86 if you need help.)

Item	Transparent/Translucent/Opaque
frosted glass	
water	
metal	
brick	
wax paper	
clear glass	

2. List the inner planets in order from the sun (nearest to farthest).

   A) _____

   B) _____

   C) _____

   D) _____

3. A rake is an example of which type of simple machine?

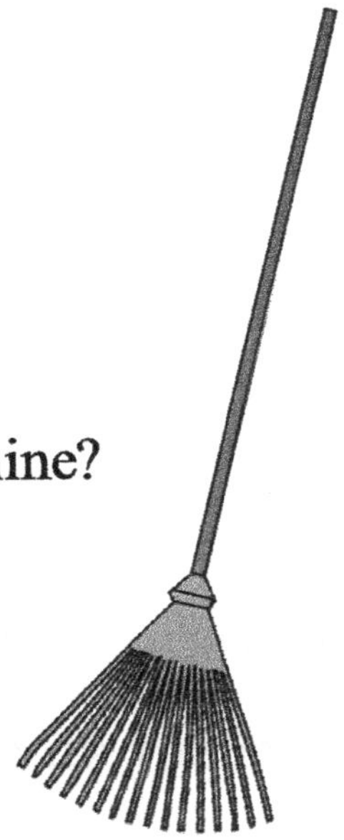

   A) wedge
   B) lever
   C) pulley

4. _____ is a type of force that stops things or slows things down.

      Matter   Friction   Sound   Gravity

5. Which layer of the Earth is below the crust? Magma forms here.

      mantle   inner core   outer core

6. A vacuum cleaner uses what kind of electricity?

      static electricity   current electricity

7. Name two ways seeds are spread.

   _____

   _____

8. Which of these has the greatest mass?

   A) a strand of hair
   B) a cotton ball
   C) a quarter
   D) a toothpick

9. A barometer measures _____.

      temperature   mass   wind speed   air pressure

10. Which animal group has moist skin and lives near water?

      reptile   bird   amphibian   mammal

Lesson #121

1. List the outer planets.

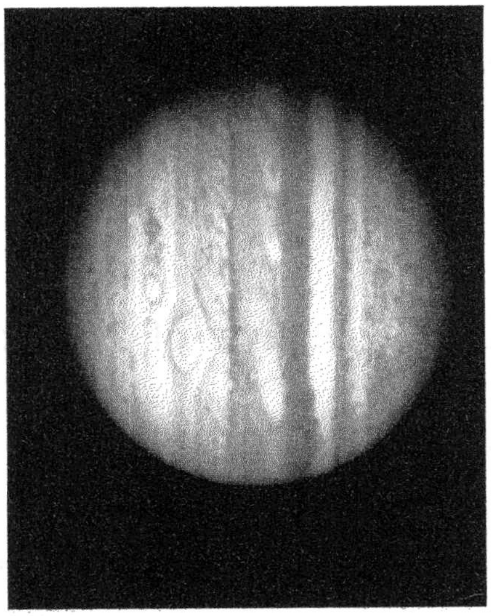

   _____

   _____

   _____

   _____

2. Put each word in the correct column of the graphic organizer.

   paint   glue   pen
   juice   oxygen   fork

Solid	Liquid	Gas

3. _____ is the decayed remains of plants and animals.

   Topsoil   Humus   Erosion   Subsoil

4. Which of these is **not** a basic need of animals?

   wings   shelter   food   water   air

5 – 6. Label the diagram using the words below.

water   sunlight   carbon dioxide   oxygen

A) _____

B) _____

C) _____

D) _____

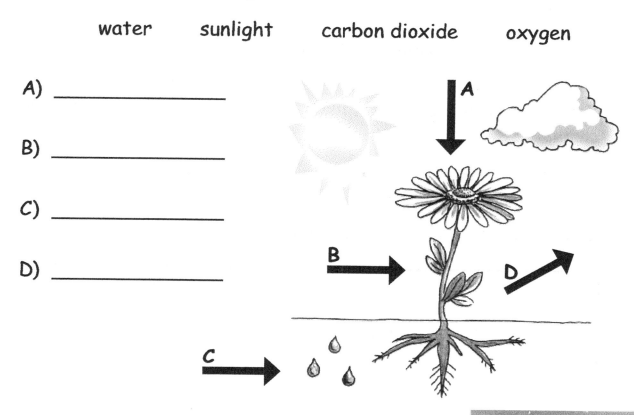

7. What does this picture represent?

evaporation   precipitation   photosynthesis

8. Materials that do not allow any light to pass through are called _____ materials.

translucent   opaque   transparent

9. Which of these is also known as a "shooting star?"

comet   asteroid   moon   meteor

10. Which of these can make their own food?

carnivores   omnivores   producers   consumers

Simple Solutions© Science   Level 4

Lesson #122

1. Which state of matter does not have a definite shape or take up a definite amount of space?

    solid   liquid   gas

2. Which organism would most likely be the final consumer in this food chain?

    fish   shark   plankton   seal

3 – 4. Draw the food chain above in the correct order.

5. Read each statement. If the statement describes a **physical change**, write a P on the line. If the statement describes a **chemical change**, write the letter C.

    ___ roasting a marshmallow   ___ shredding paper

    ___ boiling water   ___ baking a cake

6. A type of force that stops or slows things down is _____.

    sound   friction   electricity   gravity

7. Match each step of the scientific method to its definition.

    _____ Observe and Ask Questions   A) analyze the data you gathered

    _____ Form a Hypothesis   B) describe the steps you will take

    _____ Plan an Experiment   C) write a possible answer to a question

    _____ Conduct an Experiment   D) use your senses to gather information

    _____ Draw Conclusions   E) follow the steps of the experiment

8. Write T if the statement is true or F if it is false.

    _____ A) Work is done when a force moves an object.

    _____ B) The two main kinds of plants are vascular and nonvascular.

9. What is a hypothesis?

    an educated guess   an experiment   an observation

10. Choose the invertebrates.

    elephant   octopus   clam   tiger

    spider   giraffe   cat   butterfly

Lesson #123

1. Match each type of precipitation below with its description.

    ____ sleet   A) frozen rain; rain that falls through a layer of freezing cold air

    ____ snow   B) drops of water fall; temperatures above freezing

    ____ rain   C) round pieces of ice; rain freezes and falls to warmer part of the air

    ____ hail   D) drops of water vapor form ice crystals as they fall

2. Write whether each item is a **conductor** or an **insulator**.

Item	Conductor/Insulator
paper	
metal	
wood	
paper clip	

3. A doorstop is which type of simple machine?

    pulley   wedge   wheel-and-axle   level

4. Which vertebrate group is warm-blooded and has hair or fur?

    amphibian   bird   mammal   reptile

Simple Solutions© Science

5. Look at the words below. Decide which is an example of static electricity and which shows an example of current electricity. Write S for **static electricity** and C for **current electricity**.

   _____ lightning

   _____ using the washing machine

   _____ a balloon sticking to your sweater

   _____ using a mixer

6. In the United States, weather patterns generally move from _____ to _____.

7. Which layer of the Earth do we live on?

       outer core   mantel   inner core   crust

8. Which planet is the largest? (Check Help Pages if you're not sure.)

       Saturn   Earth   Jupiter   Neptune

9. Pluto used to be called a planet. What is it called now?

       a moon   a dwarf planet   a star   a comet

10. Put a ✓ next to each statement that is true.

    _____ A) Spiders have 8 legs and two body parts.

    _____ B) The largest biome is the desert.

Simple Solutions© Science   Level 4

Lesson #124

1 – 2. Write the name of the planet that matches its description.

Planet	Description
	windiest planet
	closest to the sun
	The Red Planet
	spins on its side
	largest planet
	known for its rings
	only planet to support life
	about the same size as Earth

3. These can be found orbiting the sun in a belt that lies between Jupiter and Mars. What are they?

      meteors   comets   asteroids

4. Which animal group has dry, scaly skin and is cold-blooded?

      reptile   bird   amphibian   mammal

5. This type of matter has a definite shape and takes up a definite amount of space. What is it?

      solid   liquid   gas

6. Name the process in which water vapor changes into liquid water.

   A) condensation
   B) precipitation
   C) evaporation

7. A cluster of sunflowers is an example of a(n) _____.

   community   population   ecosystem

8. Nonvascular means without _____.

   A) flowers
   B) stems
   C) tubes
   D) seeds

9 – 10. Next to each description, write *insect* or *spider*.

Description	Insect or Spider?
has 2 body parts	
has 6 legs	
has 3 body parts	
has 8 legs	

Lesson #125

Crossword Puzzle

Word Bank

Mercury	physical	mantle	crust
chemical	Jupiter	lever	pulley
force	gravity	friction	circuit
opaque	translucent	poles	insulator

Across

2. A force that pulls objects toward each other
5. A type of change that changes the make-up of the substance
6. The ends of a magnet
8. Material that doesn't let electricity pass through
9. The part of the Earth we walk on
12. The layer of Earth where melted rock called magma forms
13. Material that doesn't let light pass through
14. The path that electricity follows
15. A simple machine that is made up of a rope that is fitted around a fixed wheel

Down

1. A push or a pull
3. Material that lets some light pass through
4. The largest planet; made of gas
7. Type of force that stops things or slows them down
10. A bar that pivots, or turns, on a fixed point; a broom is an example
11. A type of change that doesn't change the make-up of the substance
12. The planet closest to the sun

Simple Solutions© Science Level 4

Use the word bank and the clues on the previous page to complete this puzzle.

Lesson #126

1. Which of these is the final step of the scientific method?

   A) Planning an Experiment
   B) Drawing Conclusions
   C) Observing and Asking Questions

2. Jeffery wants to measure wind speed. Which instrument should he use?

   A) anemometer
   B) barometer
   C) wind vane

3. Most plants need nutrients. These nutrients come mostly from _____.

   the air   water   the sun   the soil

4. Put each animal name under the correct heading in the chart.

   dragonfly   ant   rabbit   lobster

   catfish   cardinal   salamander   wasp

Invertebrates	Vertebrates

5. An animal that is hunted for food by other animals is called _____.

   predator   vertebrate   prey   warm-blooded

6. List the three needs of all living things.

   _____   _____

   _____

7. The path of food from one living thing to another is called _____.

   a life cycle   a food chain   photosynthesis

8. What is it called when an animal imitates the look of another animal?

   hibernation   mimicry   migration   instinct

9 – 10. Use the Venn diagram to compare and contrast the inner and outer planets. Give **one similarity** and **two differences.**

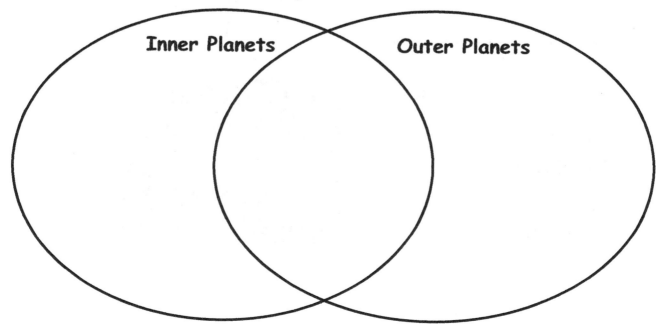

Lesson #127

1. Match each tool with its description.

   ____ barometer   A) magnify, or make something larger

   ____ thermometer   B) measure wind speed

   ____ anemometer   C) measure how hot or cold something is

   ____ hand lens   D) measure air pressure

2. A tremor or shaking of the earth's surface usually caused by movement of rock in the crust is a(n) _____.

   tornado   volcano   landslide   earthquake

3. Which type of vertebrate begins its life breathing with gills and eventually breathes with lungs?

   mammal   reptile   amphibian   fish

4. Which stage of metamorphosis is shown?

   _____

5. This is the windiest planet.

   A) Venus
   B) Neptune
   C) Uranus
   D) Earth

Look at the weather map and use it to answer the questions below.

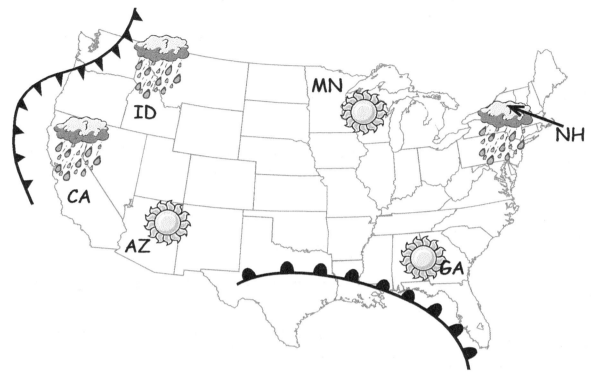

6. Which of these states is getting rain?

   Arizona   Georgia   Idaho   Minnesota

7. Which part of the country is getting a cold front?

   the northwest   the southeast

8. Circle the symbol for warm front on the map.

9. Glass and water are examples of _____ materials.

   A) translucent   B) opaque   C) transparent

10. Put a ✓ next to each statement that is true about the food chain.

    _____ A) All animals depend on plants for their energy.

    _____ B) The sun is always the first link in any food chain.

Lesson #128

1. _____ is anything that takes up space.

   Matter   Friction   Sound   Gravity

2. Match the layers of the Earth with their definition.

   ____ outer core   A) the layer below the crust; where magma forms

   ____ crust   B) this layer is solid; it is almost as hot as the sun

   ____ inner core   C) under the mantle; this layer is liquid

   ____ mantle   D) the thin outer layer; this is made up of the continents and oceans

3. The chart shows two categories. List each organism in the correct category.

   corn plant   black bear   grass   tulip   angelfish   hawk

Producer	Consumer

4. (Insects / Spiders) are the largest group of invertebrates.

Simple Solutions© Science   Level 4

5. How will these magnets react to each other?

   | S   N | | N   S |

       attract   repel

6. Using a hairdryer is an example of _____.

       current electricity   static electricity

7. Which type of cloud is shown?

   A) cirrus   C) cumulonimbus

   B) cumulus   D) stratus

8. Something that lets electricity pass through easily is called a(n) _____.

       conductor   circuit   insulator   static

9. Which type of simple machine is made of a small axle and a wheel that turns together?

   A) inclined plane

   B) lever

   C) pulley

   D) wheel-and-axle

10. The function of a plant's roots are to take in water and nutrients and _____.

        make seeds   anchor the plant   make food   produce fruit

Lesson #129

1. Study each list of words and decide what all the words have in common. Then write a category name at the top of each list.

A)	B)	C)
spider	food	crust
ant	water	mantle
lobster	air	outer core
jellyfish	shelter	inner core

2. What three things are needed for photosynthesis to occur?

    A) sunlight, sugar, and oxygen
    B) carbon dioxide, sugar, and sunlight
    C) carbon dioxide, water, and sunlight

3. List an example of a physical change.

    _____

4. What does this symbol mean?

    _____

5. The green substance inside leaves is called _____.

    sugar   loam   chlorophyll   silt

6. Materials that let no light pass through are called _____.

   transparent   opaque   translucent

7. Which of these has the **smallest** mass?

   A) a microwave
   B) a bicycle
   C) a cell phone
   D) a television

8. List water's three states.

   _____

   _____

   _____

9. What word means *the colors and patterns an animal uses to disguise itself*?

   A) instinct
   B) camouflage
   C) mimicry

10. List two traits of fish.

    _____

    _____

Lesson #130

1 – 2. Put each word in the correct column of the graphic organizer.

    liquid soap   hail   carbon dioxide

    water vapor   watch   milk

Solid	Liquid	Gas

3. List the outer planets.

   _____

   _____

   _____

   _____

4. Water evaporating is an example of a _____.

   A) solid changing to a liquid
   B) gas changing to a solid
   C) liquid changing to a gas
   D) gas changing to a liquid

5. A shovel is an example of which type of simple machine?

   lever   inclined plane   pulley   wedge

6. Which of these make their own food?

   A) consumers
   B) herbivores
   C) omnivores
   D) producers

7. Write T if the statement is true or F if it is false.

   _____ Nonvascular plants do not have tubes to carry water and food to the parts of the plant.

8. Which ecosystem is the largest in the world?

   tundra   taiga   desert   rainforest

9. List two things that can be recycled.

   _____

   _____

10. Which of these is **not** a type of soil?

    A) sand   C) clay
    B) loam   D) sediment

Lesson #131

1. Materials that allow some light to pass through are called _____ materials.

    translucent   opaque   transparent

2. When light hits an object and bounces off the surface, this bouncing off is called _____.

    photosynthesis   vibration   reflection   electricity

3. Study each list of words and decide what all the words have in common. Then write a category name at the top of each list.

A)	B)	C)
Jupiter	snake	wood
Saturn	lizard	paper
Uranus	alligator	cardboard
Neptune	turtle	brick

4. Plants make their own food through a process called _____.

    A) metamorphosis
    B) erosion
    C) weathering
    D) photosynthesis

5. In a (physical / chemical) change, the composition or make-up of the substance does not change.

6. Which clouds are associated with powerful thunderstorms?

      cumulus   cumulonimbus   stratus   cirrus

7. Name the five vertebrate groups.

    _____

    _____

    _____

    _____

    _____

8. In which stage of metamorphosis is the organism a caterpillar?

      larva   egg   pupa

9. The movement of water through the environment is called the _____.

      precipitation   evaporation   water cycle

10. List the inner planets.

    _____

    _____

    _____

    _____

Lesson #132

1. Look at the words below. Decide which is an example of static electricity and which is an example of current electricity. Write S for **static electricity** and C for **current electricity**.

   _____ A) touching an object and getting a shock

   _____ B) using a hair dryer

   _____ C) socks sticking together out of the dryer

   _____ D) using a toaster

2. Which is an example of a mammal?

   robin   salamander   whale   perch   turtle

3. This type of matter has no definite shape and does not take up a definite amount of space. What is it?

   A) solid
   B) liquid
   C) gas

4. A force that pulls objects toward each other is called _____.

   weathering   friction   gravity   reflection

5. If you put two like poles of a magnet together, what will happen?

   A) They will attract.
   B) They will repel.
   C) Nothing will happen.

6. A grizzly bear eats nuts, berries, and roots. It also eats rodents and even moose. How would you classify the grizzly bear?

   producer   carnivore   omnivore   herbivore

7. In which of these ecosystems would you find a toucan, leaf cutter ants, a spider monkey, and a poison arrow frog?

   A) deciduous forest   C) rainforest
   B) tundra   D) taiga

8. When two different large air masses meet, a _____ is formed.

   cloud   hurricane   front   fault

Use the water cycle diagram to answer the questions below.

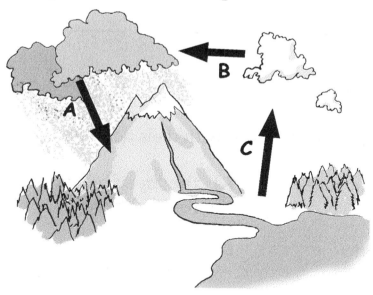

9. What process is occurring at **A**?

   precipitation   evaporation   condensation

10. What process is occurring at **B**?

    precipitation   evaporation   condensation

Lesson #133

1. The frozen layer of subsoil is called _____.

    snow   sleet   hail   permafrost

2. What do plants give off during photosynthesis?

    carbon dioxide   sugar   water   oxygen

3. Which step of the scientific method comes right after **Forming a Hypothesis**?

    _____

4. Animals have four basic needs. List them.

    _____   _____

    _____   _____

5. Match the layers of the Earth with their definition.

    ____ outer core   A) the layer below the crust; magma is formed here

    ____ crust   B) this layer is solid; it is almost as hot as the sun

    ____ inner core   C) under the mantle; this layer is liquid

    ____ mantle   D) the thin outer layer; this is made up of the continents and oceans

6. Write the name of the planet that matches its description.

Planet	Description
	spins on its side
	closest to the sun
	only planet to support life
	largest planet
	windiest planet
	The Red Planet

7. A ramp is an example of which type of simple machine?

    lever   inclined plane   pulley   screw

8. Both plants and animals need air to survive. What part of the air do animals use and what part of the air do plants use?

plants use _____

animals use _____

9. A force is a _____ or a _____.

10. What is a large body of air called?

    wind   thunderstorm   cold front   air mass

Simple Solutions© Science Level 4

Lesson #134

1 – 2. Read each type of change below. Write each type of change under the correct heading in the chart.

- boiling water
- cutting hair
- coloring a paper
- lighting a match
- burning wood
- sawing a log

Physical change	Chemical change

3. Write whether each item is **transparent, translucent,** or **opaque**.

Item	Transparent/Translucent/Opaque
water	
clear plastic	
cardboard	
wax paper	
frosted glass	

4. Something that doesn't let electricity pass through easily is called a(n) _____.

    conductor   circuit   insulator   static

5. When amphibians are born, they are most like _____.

    birds   mammals   fish

6. Explain what makes your answer to item 5 correct.

_____

_____

7. Which type of cloud is shown?

_____

8. Draw a symbol for a cold front in the box.

9. (Hypothesis / Experiment) means *an educated guess.*

10. Animals without backbones are called _____.

Lesson #135

1. Which of the following planets is an outer planet?

    A) Venus

    B) Mercury

    C) Mars

    D) Uranus

2. A force that would slow down or stop the motion of a bicycle is _____.

    gravity   friction   electricity   wind

3. Study each list of words and decide what all the words have in common. Then write a category name at the top of each list.

A)	B)	C)
robin	paper clip	Mercury
cardinal	copper wire	Venus
owl	aluminum foil	Mars
penguin	cans	Earth

4. The job of a plant's leaves is to _____.

    support the plant   get nutrients from the soil   make food

5. Which are **nonliving** parts of the environment?

   soil   sheep   fog   sponge   root

6. What does this instrument measure?

   A) wind direction   C) air pressure

   B) wind speed   D) temperature

7. Match each type of precipitation below with its description in the table.

   hail   snow   rain   sleet

Type of Precipitation	Description
	frozen rain; rain that falls through a layer of freezing cold air
	drops of water vapor form ice crystals as they fall
	drops of water fall; temperatures above freezing
	round pieces of ice; rain freezes and falls to a warmer part of the air

8. What is it called when an animal imitates the look of another animal?

   hibernation   mimicry   migration   instinct

9. Which planet is the largest?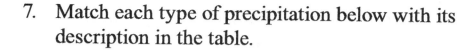

10. The largest group of invertebrates is _____.

Lesson #136

1. Which former planet is now called a "dwarf planet"?

   _____

2. When substances are changed into different substances, this is an example of a _____ change.

3. Put each planet under the correct heading in the chart.

   Mars   Uranus   Earth   Jupiter
   Neptune   Venus   Mercury   Saturn

Inner Planets	Outer Planets

4. Draw the symbol that means *recycle*.

5. Which of these is not a layer of Earth?

   inner core   mantle   clay   crust

6. What is the process called when water vapor turns into liquid water?

   A) condensation
   B) precipitation
   C) evaporation

7. What provides the energy for photosynthesis to take place?

   A) carbon dioxide
   B) oxygen
   C) sunlight
   D) water

8. This tool is used to magnify, or make something look larger. What is it called?

   _____

Use the food chain below to answer the next two questions.

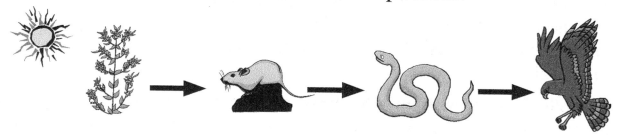

9. Which population would most likely decrease if the number of mice greatly increased?

       snakes   hawks   mice   green plants

10. Which is the producer in the above food chain?

    _____

Lesson #137

Rachel Carson: Protector of Nature (1907 – 1964)

Rachel Carson was an American scientist and writer who loved nature. She wrote books about the sea and warned people about poisons that hurt the earth. When she was young, Rachel spent a lot of time outside studying nature with her mom. Rachel loved to collect living things, and her mom taught her to treat them with care. Rachel had so many questions. But sometimes, Rachel's mother didn't answer Rachel's questions. Instead, she encouraged her daughter to research for herself. Knowing how to find answers and think things through helped Rachel Carson to become a good scientist.

One day, Rachel learned that many birds had died after being exposed to a pesticide called DDT. This pesticide was a poison used to kill insects that eat crops. The poisonous DDT was not only used on plants; it was often sprayed over animals, water, and even children who were playing outdoors. Rachel Carson began to study the effects of DDT. Then she published a book called, *Silent Spring*. It told about how pesticides hurt the earth and its inhabitants. The book made people angry. People who created and manufactured DDT called Rachel Carson a liar. Farmers didn't know what to use to stop insects from eating their crops.

Finally, President Kennedy chose a team of experts to study DDT. The study group reported that Rachel Carson was right; DDT was doing more harm than good. Soon, it became illegal to use DDT, and scientists looked for ways to repair the damage that the poison had done. Rachel Carson is remembered as a person who taught the world to pay attention and protect nature.

1. Which pesticide was the subject of Rachel Carson's book, *Silent Spring*?

   _____

Simple Solutions© Science   Level 4

2 – 3. A team of experts studied DDT at President Kennedy's request. What did they find? Answer in complete sentences.

_____

_____

_____

4. Vertebrates that are cold-blooded with dry, scaly skin are _____.

5. The by-product of photosynthesis is (chlorophyll / oxygen).

6. These are made of ice, rock, and frozen gases and have very long tails.

   meteors   planets   comets   asteroids

7. Write T if the statement is true or F if it is false.

   _____ The inner core is almost as hot as the sun.

8. A broom is an example of which type of simple machine?

   lever   pulley   wedge   inclined plane

9. To which group do salamanders and frogs belong?

   reptiles   amphibians   mammals   vertebrates

10. A group of evergreen trees is an example of a(n) _____.

    community   population   ecosystem

Lesson #138

Anna Comstock: Student of Nature (1854 – 1930)

When she was fourteen, Anna was asked to step in for her teacher, who had taken a few months off. After teaching the students multiplication and spelling, Anna would lead the children through the woods, teaching them the names of the wildflowers and bugs. Her mother had taught her how to do that when she was just a little girl. By seventeen, Anna was teaching full time.

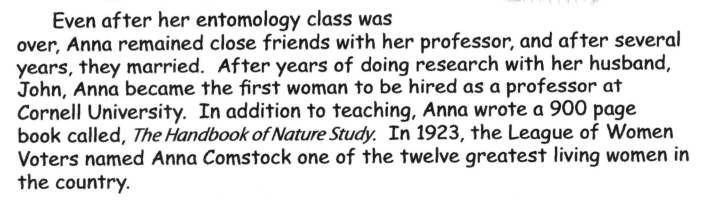

A visitor to her village saw Anna's potential and recommended she attend Cornell University, which she soon did. Anna's favorite class was entomology, the study of insects; the class was often held outdoors. Anna was never bored learning about all of the six-legged creatures.

Even after her entomology class was over, Anna remained close friends with her professor, and after several years, they married. After years of doing research with her husband, John, Anna became the first woman to be hired as a professor at Cornell University. In addition to teaching, Anna wrote a 900 page book called, *The Handbook of Nature Study*. In 1923, the League of Women Voters named Anna Comstock one of the twelve greatest living women in the country.

1. What was Anna's favorite class in college?

   _____

2. Anna was the first woman to do what?

   _____

3. What was the name of Anna's book?

   _____

4. What do you call the place where an organism lives in an ecosystem?

    its prey   its instinct   its habitat

5. Which is **not** a characteristic of birds?

    A) They lay eggs.   C) They are cold-blooded.
    B) They have feathers.   D) They have 2 feet.

6. How long does it take the moon to go through its phases?

    about 22 days   about 30 days   about 15 days

7. Match each word with its definition.

    _____ transparent   A) lets most light pass through

    _____ translucent   B) lets some light pass through

    _____ opaque   C) lets no light pass through

8. Which planet is closest to the sun?

    Venus   Mars   Mercury   Saturn

9. Which kind of electricity causes socks to stick together just out of the dryer?

    static electricity   current electricity

10. What three things are needed for photosynthesis to occur?

    A) carbon dioxide, water, and sunlight
    B) carbon dioxide, sugar, and sunlight
    C) sunlight, sugar, and oxygen

Lesson #139

Norman Borlaug: Father of the Green Revolution (1914 – 2009)

Norman Borlaug was born in 1914 on a farm near Cresco, Iowa. He died in 2009 at the age of 95. Norman Borlaug was a distinguished professor at Texas A & M University. Norman won the Nobel Peace Prize in 1970 for his role in helping to stop global famine (starvation) during the second half of the 20th century; he helped save over a billion lives. Because of his discoveries, world food production more than doubled between 1960 and 1990, especially in Pakistan and India, the two nations that benefited most from the new crop varieties discovered by Borlaug. These important changes in the agriculture of poor countries gave him the title, "Father of the Green Revolution."

Borlaug began this work in Mexico at the end of World War II. He used creative new breeding techniques to produce varieties of wheat that were resistant to disease and produced much more grain than traditional methods had in the past. Borlaug was known not only for his skill in breeding plants, but also for his willingness to work in the fields, rather than to have his assistants do all the hard work.

In July 2007, Norman Borlaug received the Congressional Gold Medal, the highest honor that can be given to a civilian by Congress.

1. Name two awards that Norman Borlaug won during his lifetime.

   _____

   _____

2. Where did his work of breeding plants begin?

   _____

3. An organism with a backbone is called a _____.

4. Anything that takes up space is called _____.

        friction   gravity   matter   force

5 – 6. Put each word in the correct column of the graphic organizer.

    juice   fog   stuffed animal
    carbon dioxide   glue   oxygen

Solid	Liquid	Gas

7. What layer of soil lies just below topsoil?

      subsoil   bedrock   humus

8. An animal that hunts another animal for food is called _____.

    prey   predator   herbivore   producer

9. Which phase of the moon is shown?

_____

10. Lighting a match is an example of a (physical / chemical) change.

Simple Solutions© Science Level 4

Lesson #140

1. Match each tool below with its description in the table.

    microscope   hand lens   balance   thermometer

Tool	Description
	This tool measures the mass of objects.
	This tool measures how hot or cold something is.
	This hand-held tool is used to magnify, or make something look larger.
	This tool is helpful to see objects that are too small to see with your eyes alone.

2. In which step of the scientific method do you look at all of the information you have collected to see if the results support your hypothesis?

    A) Conducting an Experiment
    B) Observing and Asking Questions
    C) Drawing Conclusions
    D) Forming a Hypothesis

3. Which word means *an educated guess*?

    hypothesis   investigation   observation

4. Which planet is the farthest from the sun?

    Jupiter   Uranus   Neptune   Saturn

5. A force is a _____ or a _____.

Simple Solutions© Science   Level 4

6. Match each animal group to its definition.

   _____ Fish   A) has hair or fur; feeds young with milk

   _____ Reptile   B) two legs, wings, and feathers

   _____ Bird   C) moist-skin, lives near water

   _____ Amphibian   D) dry, scaly animal that lays eggs

   _____ Mammal   E) lives whole life in water, breathes with gills

7. A plant produces _____ through photosynthesis and then uses it for energy.

   oxygen   sugar   water   carbon dioxide

8. What do you call an organism that makes its own food?

   a producer   a consumer   a food chain

9. The path that electricity follows is called a _____.

   circuit   current   fuse   conductor

10. Which sequence shows forms of water from the coldest to the hottest temperatures?

    A) ice, water vapor, liquid water
    B) water vapor, liquid water, ice
    C) liquid water, water vapor, ice
    D) ice, liquid water, water vapor

Level 4

Science

Help Pages

Help Pages
Glossary

A

air mass — a large body of air (L #25)

amphibian — a vertebrate that lives part of its life in the water and part on land (L #14)

anemometer — an instrument that measures wind speed (L #5)

asteroid — an object in space that is made up of rock and metal; orbits the sun in a belt between Jupiter and Mars (L #105)

attract — to pull toward (L #92)

B

balance — a tool that is used to measure the mass of objects (L #8)

barometer — a tool that measures air pressure (L #6)

bedrock — the final layer of soil; it lies below the subsoil (L #39)

bird — a vertebrate that has feathers, wings, and two legs (L #15)

C

carnivore — an animal that only eats other animals (L #22)

chemical change — a change in which the make-up of a substance is changed into one or more new substances (L #69)

chlorophyll — the green substance inside leaves (L #13)

circuit — the path that electricity follows (L #95)

Simple Solutions© Science Level 4

Help Pages
Glossary

cirrus — the highest clouds in the sky; they look like wisps of hair (L #30)

clay — a very fine-grained soil (L #41)

cold front — cold, dense air; appears on a weather map as a blue line with triangles on it (L #27)

cold-blooded — describes an animal whose body temperature changes with its surroundings; it cannot maintain its own stable temperature (L #15)

comet — an object in space made of ice, rock, and frozen gases; it has a very long tail (L #105)

community — all of the populations that live in an ecosystem at the same time (L #56)

condensation — the process in which water vapor (gas) turns into liquid water (L #50)

conductor — something that lets electricity pass through it easily (L #94)

conifer — plant that makes seeds inside cones (L #71)

consumer — an organism that eats other living things in order to get energy (L #20)

crust — the thin outer layer of the Earth (L #77)

cumulonimbus — describes clouds that look more like tall towers than regular cumulus clouds; associated with powerful thunderstorms (L #30)

cumulus — a name for clouds that look puffy and can bring storms (L #30)

current electricity — electricity that moves through wires (L #93)

Help Pages
Glossary

D

deciduous — describes a plant that sheds its leaves at the end of a growing season

desert — a very dry ecosystem that gets less than 10 inches of rain per year

E

earthquake — a tremor or shaking of the earth's surface usually caused by movement of rock in the crust (L #34)

ecosystem — all of the living and non-living things that interact with each other (L #54)

erosion — the process by which the surface of the Earth gets worn down (L #43)

evaporation — the process by which liquid water changes to water vapor or a gas (L #49)

experiment — a test that is done to see if the hypothesis is correct or not (L #2)

F

fault — a break in Earth's crust, where rock moves (L #34)

fish — a vertebrate group that lives in the water and is covered with scales that protect it and help it to swim (L #15)

food chain — the path of food from one living thing to another (L #20)

Help Pages
Glossary

force — any kind of push or pull (L #79)

forest — an ecosystem where many trees grow

fossil — the imprint or remains of something that lived long ago; a skeleton or leaf imprint (L #32)

friction — a type of force that stops things or slows them down (L #79)

front — the meeting of two different large air masses (L #27)

fulcrum — the fixed point on a lever (L #97)

fungi — organisms that cannot make their own food; absorb nutrients from other living things or from the remains of living things (L #10)

G

gas — a state of matter; a gas does not have a definite shape or take up a definite amount of space (L #66)

glacier — a large body of ice that moves slowly down a slope or valley (L #35)

grassland — an ecosystem that is usually dry and flat

gravity — a force that pulls objects toward each other (L #80)

H

hand lens — a tool that is used to magnify or make something look larger (L #12)

herbivore — an animal that eats only plants (L #22)

horizon — another word for soil layers (L #39)

Help Pages
Glossary

humus — the decayed remains of plants and animals (L #39)

hurricane — a large tropical storm with wind speeds of 74 mph or more; it forms over warm, tropical waters (L #52)

hypothesis — an educated guess, or a possible answer to a question (L #1)

I

ice sheet — a type of glacier found in Greenland and Antarctica; covered much of the Earth long ago (L #35)

inclined plane — a simple machine that makes lifting and moving things a lot easier; a ramp is an example (L #99)

inner core — the center of the Earth; the inner core is solid because there is so much pressure on it; it is almost as hot as the sun (L #77)

insulator — something that does not let electricity flow through it easily (L #94)

invertebrate — an animal without a backbone (L #17)

investigation — a study a scientist conducts when there is a problem or a question to be answered (L #1)

L

landslide — the sliding of a mass of loosened rock or earth down a hillside or slope (L #36)

larva — the second stage of metamorphosis (L #23)

Help Pages
Glossary

leaf — the part where the plant makes its food (L #4)

lever — a bar that pivots, or turns, on a fixed point (L #97)

liquid — a state of matter that takes the shape of its container (L #66)

living — alive; living things need food, water, and air in order to live (L #1)

loam — a mixture of all of the types of soil: humus, clay, silt, and sand; great for growing crops (L #41)

M

magnet — a piece of metal that attracts iron and steel (L #92)

mammal — a vertebrate that has hair or fur, is warm-blooded, and gives birth to live young (L #14)

mantle — the layer of the Earth below the crust; where magma is formed (L #77)

matter — anything that takes up space (L #65)

metamorphosis — the series of changes in appearance from birth to adulthood in some animals (L #23)

meteor — a shooting star; meteors are usually very tiny (L #105)

microscope — a tool scientists use to magnify an object; helps to view objects which cannot be seen with only the eyes

mimicry — imitating the look of another animal (L #46)

moon phases — different shapes of the moon seen at different times, having to do with the position of the moon and the sun (L #101)

Help Pages
Glossary

N

nonliving — not alive (L #1)

nonvascular — describes a plant that absorbs water through its surfaces like a sponge; there are no tubes to transport water and nutrients (L #8)

O

omnivore — an animal that eats both plants and other animals (L #22)

opaque — materials that do not allow any light to pass through (L #86)

outer core — the layer of the Earth below the mantle; the outer core is liquid (L #77)

P

permafrost — a layer of frozen subsoil found in the tundra (L #59)

photosynthesis — the process by which plants use energy from sunlight to turn water and carbon dioxide into sugar called glucose; means *making things with light* (L #13)

physical change — a change that does not create some new substance (cutting paper, ice-cube melting) (L #67)

pitch — the highness or lowness of a sound (L #91)

poles — the ends on a magnet (L #92)

population — a group of organisms of the same kind (L #55)

Help Pages
Glossary

precipitation — any form of water that falls to the ground (rain, snow, sleet, or hail) (L #51)

predator — an animal that hunts another animal for food (L #19)

prey — an animal that is hunted by another animal for food (L #19)

producer — an organism that makes its own food (L #20)

pulley — a simple machine that is made up of a rope that is fitted around a fixed wheel (L #98)

pupa — a stage of metamorphosis in which the organism is wrapped in a cocoon (L #23)

R

rainforest — dense forest located along the equator that gets up to 33 feet of rain per year; home to millions of plants and animals (L #58)

recycle — to conserve resources by reusing the same resource (L #61)

reflection — light bouncing off the surface of an object (L #85)

repel — to push away (L #92)

reptile — a vertebrate that has dry, scaly skin and lays eggs on land (L #15)

root — a part of a plant that takes in water and nutrients from the soil (L #4)

S

sand — a type of soil that has tiny grains of rock that can easily be seen with the eyes (L #41)

Help Pages
Glossary

scientific method — an organized plan that a scientist uses to conduct a study (L #1)

screw — a simple machine used to hold two or more objects together or to lift an object (L #101)

seed — the first stage of life for many plants; contains the food to help a new plant grow (L #4)

seismograph — an instrument that shows the movement of the Earth's surface during an earthquake (L #34)

silt — a type of soil made of tiny grains of rock (L #41)

simple machine — a tool with few or no moving parts that makes work easier (L #97)

soil — a mixture of many different things: water, air, humus, and bits of rock (L #41)

solar system — a group of objects that revolve around the sun (L #105)

solid — matter that has a definite shape and takes up a definite amount of space (L #65)

sound — a form of energy that travels in waves (L #91)

static electricity — an electric charge that builds up on an object (L #93)

stationary front — a front that stays in one place for many days (L #27)

stem — the part that holds the plant up; the stem carries water and nutrients from the roots to the leaves (L #4)

stratus — the lowest clouds in the sky; they look like a layer of clouds (L #30)

Help Pages
Glossary

subsoil — the bottom layer of soil in which the soil particles are larger and not as dark as topsoil; contains small pieces of rock (L #39)

T

taiga — a cold forest found south of the tundra and north of the deciduous forests; largest ecosystem in the world; means *forest* (L #60)

topsoil — the top layer of soil; topsoil contains a lot of humus; plants grow best there (L #39)

translucent — describes materials that let some light pass through (L #86)

transparent — describes materials that let most of the light that hits it pass through (L #86)

tundra — a cold and treeless ecosystem; found in Alaska, Greenland, Canada, Europe, and Russia (L #59)

V

vascular — describes plants that have tube-like structures that transport water from the roots to the stem and leaves (L #7)

vertebrate — an animal with a backbone (L #14)

vibrate — to move back and forth (L #91)

volcano — both the vent and the cone-like mountain left by the overflow of erupted lava, ash, and rock (L #33)

Help Pages
Glossary

W

warm front — a type of air mass that usually moves slowly and brings steady rain, rather than thunderstorms; appears on a weather map as a red line with half circles (L #27)

warm-blooded — describes an animal whose body temperature remains constant regardless of the temperature surrounding it; the animal can maintain its internal body temperature (L #15)

water cycle — the continuous movement and recycling of water throughout the environment (L #48)

weather map — a map showing what the weather is like over a specific geographic area at a specific time (L #28)

weathering — the breakdown of rocks and minerals on the Earth's surface (L #42)

wedge — a simple machine made up of two inclined planes placed back to back (L #102)

wheel-and-axle — a simple machine made of a small cylinder or an axle attached to a larger wheel (L #100)

work — something that is done when a force is used to move an object (L #96)

Help Pages

Invertebrate Fact Cards

Cockroach

Facts: Cockroaches are found everywhere in the world except in the polar regions. There are about 3,500 species of cockroach; most live in the tropics. The most likely place to see a cockroach is indoors. Cockroaches usually come out at night.

Diet: Cockroaches will eat almost anything they come across, including dead animals and animal droppings. However, plant matter makes up most of their diet. Cockroaches will also eat paper, glue, and even wallpaper in a home.

Breeding: Cockroaches lay 5 – 50 eggs which are fully grown in 3 – 6 months.

Group: Insect

Cricket

Facts: Even though crickets have wings, most do not fly. Most get around by jumping from place to place. A cricket usually lives less than 1 year. They make a song by rubbing their otherwise useless wings together.

Diet: Crickets eat green leaves, fresh seedlings, garden fresh fruits, and tomatoes; they are also known to feed on smaller insects.

Breeding: One female cricket can lay as many as 2,000 eggs in the fall, and the eggs will hatch in the spring. The baby crickets are called nymphs.

Group: Insect

Simple Solutions© Science

Help Pages

Invertebrate Fact Cards

Grasshopper

Facts: Grasshoppers can be found in most places in the world where there are plants. Many species of grasshoppers have no wings, so they must hop from place to place. Grasshoppers are good at camouflaging themselves for protection against predators. They make sounds by rubbing their back legs against their wings.

Diet: Grasshoppers eat leaves, flowers of plants, and grass.

Breeding: Grasshoppers lay between 3 and 100 eggs. When they first hatch, they look like tiny worms. They go through a process called molting or shedding their skin. Each time they shed their skin, they grow larger.

Group: Insect

Honeybee

Facts: Bees live in large colonies. They make honey, which is their source of food during the winter months. In the 1 – 7 years of her life, the queen bee will lay up to 1,500 eggs.

Diet: Bees live on nectar and pollen.

Life Span: Queen bee: 7 years, drones: 4 – 5 weeks, and workers: 8 weeks

Defense: A honeybee uses a stinger as a means of defense. The stinger remains attached to the bee's venom gland. When the bee struggles to get free from its victim, its lower abdomen, along with the stinger is torn away.

Group: Insect

Help Pages

Invertebrate Fact Cards

Housefly

Facts: One housefly can carry over a million bacteria, which can spread disease. The housefly loves garbage dumps and sewers. It cannot chew or swallow solid food, so it must suck it up in liquid form.

Diet: Houseflies consume mainly rotting flesh, fruit, and excrement.

Breeding: The housefly can lay up to 900 eggs in batches of 120 – 150 at a time. It takes only about 1 week to develop from an egg to an adult.

Group: Insect

Katydid

Facts: Katydids are mainly nocturnal. Their green color and ability to remain still during the day helps them to avoid predators. Most katydids live in trees, but some live in tall grasses or weeds. Katydids make their songs by rubbing the base of their front wings together.

Diet: Katydids eat mostly leaves, shrubs, weeds, and occasionally other insects.

Breeding: Katydids lay up to 50 eggs. Their lifespan is about one year.

Group: Insect

Help Pages

Invertebrate Fact Cards

Praying Mantis

Facts: The praying mantis can be found in many habitats, including the desert, the forest, and the grasslands. There are over 2,000 different species of mantis found around the world.

Diet: The praying mantis eats insects, spiders, and even other praying mantises.

Life Span: about 10 – 12 months

Defense: The praying mantis uses camouflage as one type of defense. Birds are a major enemy of the praying mantis. In an effort to discourage them, the mantis will strike out with its spiny forelegs.

Group: Insect

Tarantula

Facts: There are about 40 types of tarantula found in the United States. Most of these are found in the deserts of the Southwest.

Diet: Tarantulas eat beetles, spiders, grasshoppers, moths, and millipedes. Some larger species eat lizards, snakes, frogs, toads, and mice.

Life Span: about 30 years or more

Defense: The tarantula has several defenses. Some tarantulas lean back on their haunches and expose their long fangs. Other species squirt an unpleasant substance into their predator's face. A third defense comes from the hairs on its abdomen. These hairs have sharp points that can cause pain or blindness if they come into contact with the eyes or skin of an animal.

Group: Spider

Help Pages

Ecosystem Fact Cards

Deciduous Forest

Forests occupy one-third of Earth's land area. The temperate deciduous forest can be found in most of the eastern United States and a small strip of southern Ontario, Canada. The dominant plant species of this biome is broad-leaved deciduous trees. Trees such as oak, hickory, beech, maple, elm, and pine can be found in the deciduous forests.

The deciduous forest has four distinct seasons. The leaves of deciduous trees change color and fall off in the autumn and grow back in the spring. Temperate deciduous forests get between 30 and 60 inches of precipitation a year.

The animals that live in the forest are squirrels, deer, fox, rabbits, skunks, birds, raccoons, and black bears, to name a few.

deciduous forest

deer

fox

Help Pages

Ecosystem Fact Cards

Grasslands

Grasslands are dry and usually flat areas of land that are hot in the summer and cold in the winter. They get more rain and snow than deserts, but less than some other ecosystems. Food crops tend to grow well in the grasslands. Due to the hot, dry summers, a large number of fires can occur. These fires cause great changes to the grasslands.

The main plant of the grasslands is grass, but bushes and wildflowers also grow there. There are very few trees and shrubs in the grasslands.

Some animals of the grasslands include bison, coyotes, mice, rabbits, owls, hawks, and snakes. Many insects can be found in the grassland as well.

bison

wildflowers

grass

Help Pages

Ecosystem Fact Cards

Rainforest

Rainforests have tall trees, a warm climate, and lots of rain. Rainforests can be found in Australia, Africa, Asia, and Central and South America. The largest rainforest in the world is the Amazon rainforest in South America.

Rainforests cover less than 2% of the Earth's surface but are home to more than 50% of all of the plants and animals on Earth. Some animals found in the rainforest include jaguar, ocelot, gorilla, lemur, orangutan, bat, leaf cutter ant, poison arrow frog, toucan, macaw, sloth, wild boar, chameleon, and boa constrictor. However, insects are the most numerous animal in the rainforest.

Some foods that originally came from the rainforest include cashews, bananas, pineapple, coffee, tea, yams, cinnamon, cocoa, and peanuts. The rainforest gets 7 to 33 feet of rain each year. Some rainforests get an inch of rain per day.

macaw

ape

jaguar

Help Pages

Ecosystem Fact Cards

Taiga

Taiga is the Russian word for *forest*. It is the largest biome in the world. The taiga is located below the tundra biome. Winters in the taiga are very cold with lots of snow. The summers are warm, rainy, and humid. A lot of coniferous trees (evergreens) grow in the taiga. The taiga doesn't have as much plant and animal life as the deciduous forest or the rainforest, but millions of insects live there in the summertime, and birds migrate there every year.

The main seasons are winter and summer. Spring and fall are so short, it's as if they are not seasons at all. Not many plants can survive the harsh weather conditions, but there are some lichens and mosses, as well as pine trees, white spruce, Douglas fir, and hemlock that can grow there. Some animals found in the taiga include lynx, wolverine, bobcat, mink, elk, red deer, moose, and snowshoe rabbit.

evergreen trees

moose

wolverine

Help Pages

Ecosystem Fact Cards

Tundra

The word tundra means *treeless plain*. The Arctic tundra is one of the Earth's coldest, harshest, and windiest ecosystems. The Arctic tundra is located north of Greenland, Alaska, Canada, Europe, and Russia.

Because of the frozen soil, called **permafrost**, it is impossible for trees to grow there. The tundra is snow-covered for much of the year. The growing season lasts for only 50 – 60 days out of the year. The plant life consists mainly of mosses, lichen, small shrubs, and grasses. Some animals found in the tundra include Arctic foxes, caribou, lemmings, polar bears, reindeer, wolves, weasels, snowy owls, and snow geese.

polar bear   Arctic fox   caribou

Help Pages

The Scientific Method

When scientists have a problem or a question, they use an organized plan called the **scientific method** to conduct a study, called an **investigation**. There are 5 steps for planning and conducting an investigation.

1. **Observing and Asking Questions** – During this step, you use your senses to gather information. You may begin to think of some questions about what you are observing. You may also discover some things you don't know, but would like to find out more about.

2. **Forming a Hypothesis** – A hypothesis is a possible answer to one of the questions about your observations. It is a logical guess. A hypothesis can be tested to see if it is correct and should be written in a complete sentence.

3. **Planning an Experiment** – An experiment is a test that is done to see if your hypothesis is correct or not. When you plan an experiment, you need to describe the steps, list the materials you will need, identify the variables, and decide how you will gather and record your data.

4. **Conducting an Experiment** – Follow the steps of the experiment you planned in step 3. Observe carefully, and record your information accurately.

5. **Drawing Conclusions** – Look at all of the information you have collected. You can make graphs and charts to summarize the results. Write a conclusion and decide whether your hypothesis was correct. Share your results.

Help Pages

Stages of Frog Metamorphosis

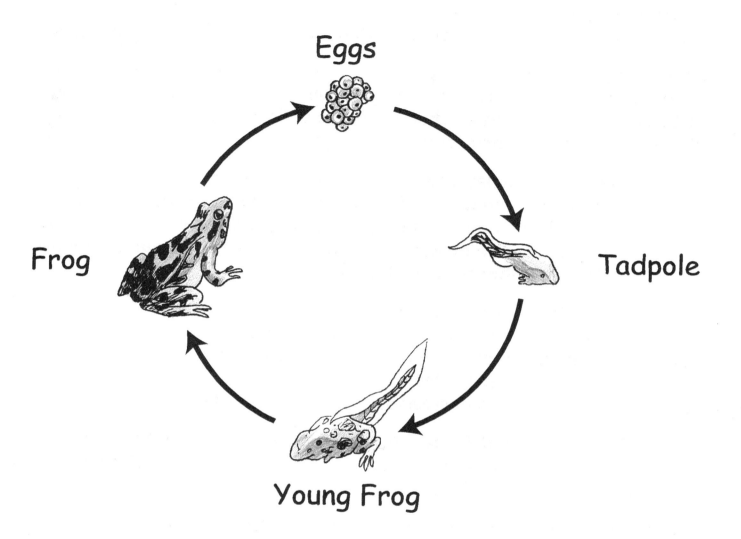

Help Pages

Simple Machines

A simple machine is a tool with few or no moving parts. It makes work easier. There are six simple machines.

Lever – A lever is a bar that pivots or turns on a fixed point. The fixed point is called the **fulcrum**. A broom is an example of a lever. Some other examples of levers are a **shovel**, a **broom**, your **arm**, and a **fishing pole**.

Pulley – A pulley is made of a rope that is fitted around a fixed wheel. It changes the direction of a force. You pull one end of the rope one way, and the other end moves in the opposite direction. Pulleys are found on **cranes, window blinds, sailboats,** and **flagpoles**.

Wheel and Axle – A wheel-and-axle is made up of a small cylinder or an axle attached to the center of a larger wheel. The wheel and axle are connected so that they turn together. Some examples of a wheel-and-axle are a **screwdriver**, a **faucet**, a **doorknob**, and a **steering wheel**.

Inclined Plane – An inclined plane makes lifting and moving things easier. A **ramp** is an example of an inclined plane.

Screw – A screw is a simple machine that you turn to hold two or more objects together or to lift an object. A screw looks like a nail with threads around it.

Wedge – A wedge is made up of two inclined planes placed back to back. You use a wedge to force two things apart or to split one thing into two things. Some examples of wedges are a **doorstop**, an **axe**, a **chisel**, and a **knife**.

Help Pages

Planets

Planet	Description
Mercury (inner)	closest planet to the sun; has craters
Venus (inner)	about the same size as Earth; covered with mountains, volcanoes
Earth (inner)	only planet to support life; most of the planet made up of water
Mars (inner)	known as the Red Planet; has huge dust storms
Jupiter (outer)	largest planet and is made up of gases; has a giant storm called the Great Red Spot
Saturn (outer)	known for its many beautiful rings; has 31 moons
Uranus (outer)	rotates on its side; has about 27 moons
Neptune (outer)	windiest planet in the solar system; blue in color

Help Pages

Science Tools

We use tools to help us observe, measure, or study objects.

An **anemometer** is an instrument that measures the speed of the wind.

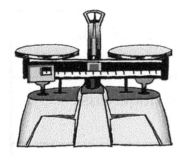

A **balance** is a tool that is used to measure the mass of objects. When you place an object on one pan and another object on the other pan, you are able to compare the objects' masses.

This tool is used to magnify an object, or make it look larger. It is called a **hand lens**.

A **microscope** is a tool used to magnify objects. Microscopes are helpful to see objects that are too small to see with your eyes alone.

This tool is used to measure how hot or cold something is. It is called a **thermometer**.

A **barometer** is used to measure air pressure.

Help Pages

Vertebrate/Invertebrate Chart

Animal Type	Image	Description
Mammal		has hair or fur; breathes with lungs; gives birth to live young; produces milk for young; warm-blooded
Reptile		has dry, scaly skin; lives near water; lays eggs; breathes with lungs; cold-blooded
Bird		has feathers, wings, and two feet; lays eggs; breathes with lungs; warm-blooded
Fish		has scales; breathes with gills; lays eggs; cold-blooded
Amphibian		lives part of its life in water and part on land; has moist skin; stays close to water; breathes with gills when born and lungs when an adult; cold-blooded
Invertebrate		has no backbone; some live on land and some in water; some crawl, some fly, and some swim